BestMasters

Mit **„BestMasters"** zeichnet Springer die besten Masterarbeiten aus, die an renommierten Hochschulen in Deutschland, Österreich und der Schweiz entstanden sind. Die mit Höchstnote ausgezeichneten Arbeiten wurden durch Gutachter zur Veröffentlichung empfohlen und behandeln aktuelle Themen aus unterschiedlichen Fachgebieten der Naturwissenschaften, Psychologie, Technik und Wirtschaftswissenschaften. Die Reihe wendet sich an Praktiker und Wissenschaftler gleichermaßen und soll insbesondere auch Nachwuchswissenschaftlern Orientierung geben.

Springer awards **„BestMasters"** to the best master's theses which have been completed at renowned Universities in Germany, Austria, and Switzerland. The studies received highest marks and were recommended for publication by supervisors. They address current issues from various fields of research in natural sciences, psychology, technology, and economics. The series addresses practitioners as well as scientists and, in particular, offers guidance for early stage researchers.

Julian Knaup

Impact of Class Assignment on Multinomial Classification Using Multi-Valued Neurons

Springer Vieweg

Julian Knaup
Lemgo, Germany

ISSN 2625-3577 ISSN 2625-3615 (electronic)
BestMasters
ISBN 978-3-658-38954-3 ISBN 978-3-658-38955-0 (eBook)
https://doi.org/10.1007/978-3-658-38955-0

Responsible Editor: Stefanie Eggert
This Springer Vieweg imprint is published by the registered company Springer Fachmedien Wiesbaden GmbH, part of Springer Nature.
The registered company address is: Abraham-Lincoln-Str. 46, 65189 Wiesbaden, Germany

Abstract

Multilayer neural networks based on multi-valued neurons (MLMVNs) have been proposed to combine the advantages of complex-valued neural networks with a plain derivative-free learning algorithm. In addition, multi-valued neurons (MVNs) offer a multi-valued threshold logic resulting in the ability to replace multiple conventional output neurons in classification tasks. Therefore, several classes can be assigned to one output neuron. This thesis introduces a novel approach to assign multiple classes on numerous MVNs in the output layer. It was found that classes that possess similarity should be allocated to the same neuron and arranged adjacent to each other on the unit circle. Since MLMVNs require input data located on the unit circle, two employed transformations are reevaluated. The min-max scaler utilizing the exponential function obtains decent results for numerical data. The 2D discrete Fourier transform restricting to the phase information was found to be unsuitable for image recognition. Even if this transformation approach could be improved, it loses key properties such as translational invariance by discarding the magnitude information. The evaluation was performed on the SDD and the Fashion MNIST dataset.

Contents

List of Abbreviations

2D DFT	Two Dimensional Discrete Fourier Transform
Adam	Adaptive Moment Estimation
AI	Artificial Intelligence
ANN	Artifical Neural Network
AutASS	Autonomous Drive Technology by Sensor Fusion for Intelligent, Simulation-based Production Facility Monitoring and Control (abbr. from German)
BP	Backpropagation
CNN	Convolutional Neural Network
CPU	Central Processing Unit
CVNN	Complex-Valued Artifical Neural Network
DT	Decision Tree
FDR	Fisher Discriminant Ratio
FN	False Negative
FP	False Positive
GD	Gradient Descent
GPU	Graphics Processing Unit
LDA	Linear Discriminant Analysis
LSTM	Long Short-Term Memory
MLMVN	Multilayer Feedforward Neural Network Based on Multi-Valued Neurons
MNIST	Modified National Institute of Standards and Technology
MSE	Mean-Square-Error
MVN	Multi-Valued Neuron
NLP	Natural Language Processing
PCA	Principal Component Analysis
ReLU	Rectified Linear Unit
RMSE	Root-Mean-Square-Error
RNN	Recurrent Neural Networks

RVNN	Real-Valued Artifical Neural Network
SDD	Sensorless Drive Diagnosis
SVD	Singular Value Decomposition
SVM	Support-Vector Machine
t-SNE	t-Distributed Stochastic Neighbor Embedding
TN	True Negative
TP	True Positive
UMAP	Uniform Manifold Approximation and Projection for Dimension Reduction
XOR	Exclusive Disjunction

List of Symbols

α	scaling factor
a	feature minimum
b	feature maximum
$\mathbb{C}$	set of complex numbers
γ, Γ	angle between y and d, angle matrix between Y and D
d, D	desired output/bisector, desired output/bisector matrix
δ, E	error, error matrix
∂	partial derivative
ε_k, E_k	sector border, set of sector borders
$\mathbb{H}$	set of quaternions
$\Im$	imaginary part
j	imaginary number
$\mathcal{L}$	loss function
M	image height
N	image width
η	learning rate
$\mathcal{N}$	gaussian/normal distribution
$\mathbb{N}$	set of natural numbers
$P()$	activation function
φ	angle of z
r	circle radius
$\mathbb{R}$	set of real numbers
$\Re$	real part
$\Sigma^{\dagger}$	pseudo inverse of the non-square matrix Σ
$\mathcal{U}$	uniform distribution
U^H	conjugate transpose of U
w_0	bias
w, W	weight, weight matrix

$\Delta w, \Delta W$	weight adjustment, weight adjustment matrix
$\tilde{w}$	updated value of w
x_0	pseudo input for multiplication with bias
x, X	input sample, input matrix
$\bar{x}$	complex-conjugated value of x
$\|\|x\|\|_p$	L_p norm of z
y, Y	neuron output, output matrix
z	weighted sum
$\|z\|$	absolute value of z

Chapter 1

Introduction

> A computer program is said to learn from experience E with respect to some task T and some performance measure P, if its performance on T, as measured by P, improves with experience E.
>
> —Tom Mitchell, Definition of machine learning [Mit97]

Given this definition, it becomes clear that the deployment of machine learning can bring tremendous benefits with its ability to adapt itself. Therefore, machine learning has recently gained popularity and importance. In particular, areas of pattern recognition and natural language processing have been revolutionized by Deep Learning [KSH12][HS97]. Although artificial neural networks do not solve arbitrary problems in a deus ex machina manner, they surpass human performance in selected areas [HZRS15]. Hence, they can be applied following the principle that all models are wrong, but some are useful.

In the past, the machine learning community has primarily focused on real-valued networks that use backpropagation to adjust their weights and biases through some form of gradient descent. However, complex-valued neural networks have the advantage of more extensive mapping space. The increased functionality due to the complex weights, biases, and activation functions leads to networks with fewer neurons, among other benefits [Aiz11]. Therefore, Trabelsi et al. [TBZ+18] provide key components such as complex batch normalization for frameworks to train deep complex-valued neural networks.

Nevertheless, these networks suffer from the same drawbacks as real-valued neural networks based on gradient descent, i.e., they have to use differentiable activation functions and compute, or at least estimate, the gradient. On the other hand, there are networks based on complex-valued neurons using a simpler learning algorithm since it is derivative-free [Aiz11]. It is a direct error correction approach using phase information but omitting

J. Knaup, *Impact of Class Assignment on Multinomial Classification Using Multi-Valued Neurons*, BestMasters, https://doi.org/10.1007/978-3-658-38955-0_1

magnitude information. These neurons are referred to as multi-valued neurons (MVNs). The attribute multi-valued means in this context that the neurons do not only classify in a binary way but can distinguish between several classes. Therefore, MVNs can replace multiple output neurons. Moreover, due to their complex-valued properties, MVNs also have higher functionality than real-valued neurons, which can be increased by interconnecting multiple MVNs. The resulting multilayer neural networks based on multi-valued neurons (MLMVNs) thus require fewer neurons than real-valued networks.

However, when reducing the number of output neurons below the number of classes, it is necessary for single neurons to be able to indicate several classes. This is currently accomplished by a single output neuron that processes all classes. However, when using multiple MVNs that process several classes, the question arises how classes are assigned to the output neurons. It has to be clarified whether this kind of classification is generally preferable and, if so, what impact the class assignment has. It has to be investigated on which foundation class assignments have to be executed.

1.1 Research Objectives

Within this master thesis, first, an extension of an existing framework for MLMVNs will be accomplished to allow experimental verification of the advantages and disadvantages of MLMVNs on benchmark datasets. Furthermore, MLMVNs currently do not utilize multiple output neurons to exploit their multi-class classification capabilities. Therefore, it will be investigated to what extent multiple multi-valued output neurons with multiple assigned classes are able to classify performantly. In particular, the impact on the classification rate regarding the allocation of certain classes to the MVNs and their arrangement on the MVN will be investigated. Therefore, the following work is performed:

- a comprehensive study of the MVNs resp. MLMVNs, as well as further preliminaries
- a state of the art analysis with respect to classifications by real-valued and complex-valued neural networks
- a development of a concept for investigating the impact of class assignments on the classification rate utilizing MVNs
- an evaluation of the concept on two datasets

1.2 Scope of Work

In this thesis, a recommendation for the assignment of classes to MVNs in the classification process will be developed. For this purpose, the entire chain of machine learning from preprocessing over training to classification is described in the following. Here, both a tabular numeric dataset and an image dataset are preprocessed and evaluated using existing methods. Tabular numeric denotes a numeric dataset containing different features in its columns, and the rows represent several samples. It will be referred to as numeric dataset in the following. On the other hand, an image dataset contains pixel values that have a discrete two-dimensional position within a picture. The goal is not to achieve the best preprocessing and features but to critically evaluate existing methods. The main focus is to investigate the impact of class assignments within the classification process.

1.3 Structure of the Work

This master thesis is divided into six chapters. The introduction as well as the objectives and their scope, have been emphasized in this chapter. In Chapter 2, the reader is familiarized with the necessary preliminaries of neural networks, particularly those dealing with MVNs and MLMVNs. Afterwards, in Chapter 3, a state of the art analysis is carried out with respect to key components in the classification. This includes both real-valued and complex-valued neural networks. The novel approach of how class assignments are applied to MVNs is addressed in Chapter 4. For this purpose, existing methods for preprocessing are leveraged, and the training model is presented. The concept evaluation on two datasets follows in Chapter 5, where not only the datasets are presented, but implementation details are given, and the results are discussed. Chapter 6 completes the thesis with a conclusion and an outlook.

Chapter 2

Preliminaries

This chapter provides the reader with the necessary preliminaries to understand the following chapters. First, the basic concepts of machine learning are introduced in Chapter 2.1, and then the computational models of artificial neural networks are explored in more detail in Chapter 2.2. For this purpose, a brief review of the history of artificial neural networks is given, and their architecture and training are discussed. Afterwards, the focus is dedicated to multi-valued neurons and their interconnection to multilayer networks. Subsequently, a closer look at loss functions is given in Chapter 2.3 and the most crucial information from this chapter is summarized in Chapter 2.4.

2.1 Machine Learning Basics

To revisit Mitchell's quote, "A computer program is said to learn from **experience E** with respect to some **task T** and some **performance measure P**, if its performance on T, as measured by P, improves with experience E" from Chapter 1, the machine learning basics are divided into the points of experience, task, and performance measure.

Experience E

The experiences E are findings obtained from the respective datasets. Thereby machine learning algorithms can be roughly divided into three different methodologies, supervised learning, unsupervised learning and reinforcement learning. The latter is not essential for further considerations.

Supervised learning refers to learning with samples x that are assigned to a target value y. In the learning process, the algorithm computes an error between the predicted value and the corresponding target value and performs an adjustment to the model based on

J. Knaup, *Impact of Class Assignment on Multinomial Classification Using Multi-Valued Neurons*, BestMasters, https://doi.org/10.1007/978-3-658-38955-0_2

this error. The goal is to build a model that predicts the target value y for an unknown sample x, i.e., to determine an estimate of $p(y|x)$ [GBC16]. Well-known representatives of this learning methodology are support-vector-machines (SVM), decision trees (DT) and artificial neural networks (ANN).
In unsupervised learning, sample x lacks an associated target value y. The algorithm tries to learn the structure of the dataset by itself. In contrast to supervised learning, it does not try to find $p(y|x)$ but the probability distribution $p(x)$ [GBC16]. Typical applications are dimensionality reduction, such as Principal Component Analysis (PCA), and clustering algorithms, for instance, k-Means.

Task T

The task T decides how a sample x is processed in the system. Typical machine learning tasks are regression, classification, denoising, and clustering, but there are many more. In this work, the scope is limited to clustering and classification tasks.
Clustering is an unsupervised learning method. The algorithm tries to find similarities in unlabeled data and group them. Such a procedure is suitable for the general evaluation of datasets or is used in the context of anomaly detections.
Classification is a supervised learning method. In a k-valued classification problem, $y \in \{0, 1, ..., k-1\}$. The algorithm tries to find a function f that satisfies $f(x) = y$ or outputs a probability distribution over the class membership. Image recognition is a prominent example of classification tasks in this context.

Performance Measure P

To quantify the performance of an ML model, it requires a measure P. This measure is dependent on task T, which is why the following is restricted to measures from the area of classification. Since only balanced datasets are used, and there are no error cases that are more severe than others, no advanced methods are needed. Thus, in this case, accuracy and the confusion matrix would be sufficient to compare results. Accuracy (2.1) expresses the ratio of correctly classified examples to the total set of samples [Dal18], and the confusion matrix (Fig. 2.1) additionally shows us graphically where the error cases are placed. However, since accuracy is a global metric and a class-related value is misleadingly high, as misclassified samples that do not belong to the actual class are considered as TN, the F1-score [Dal18] is used to evaluate the results of the individual classes. The F1-score (2.4) is composed of precision (2.2) and recall (2.3) [Dal18] and can be determined from the confusion matrix:

$$Accuracy = \frac{TP + TN}{TP + TN + FP + FN}, \tag{2.1}$$

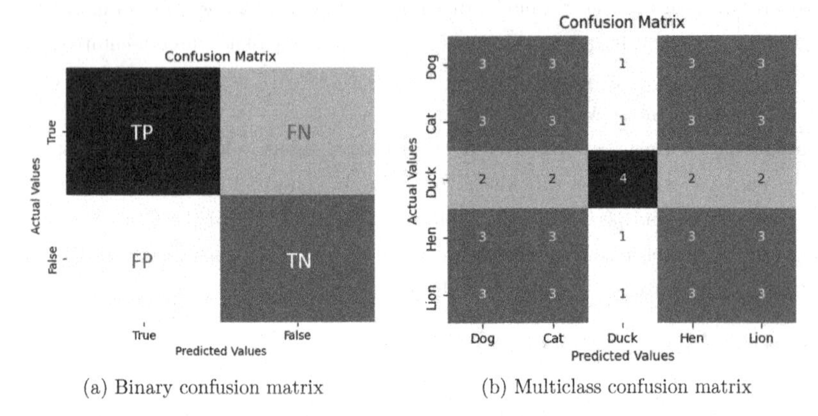

(a) Binary confusion matrix

(b) Multiclass confusion matrix

Figure 2.1: Confusion Matrix

$$Precision_K = \frac{TP_K}{TP_K + FP_K}, \tag{2.2}$$

$$Recall_K = \frac{TP_K}{TP_K + FN_K}, \tag{2.3}$$

$$F1 - Score_K = 2 \cdot \frac{Precision_K \cdot Recall_K}{Precision_K + Recall_K} = \frac{TP_K}{TP_K + 0.5(FP_K + FN_K)}. \tag{2.4}$$

For the single class evaluation, the definition for the binary case remains; the index K is added for the respective class, given that Micro and Macro F1-score, as used in multiclass tasks, are also global metrics [SL09]. Figure 2.1a is expanded to 2.1b such that, using the duck as an example, the value four is the TP, the row (the twos) forms the FN, and the column (the ones) forms the FP.

2.2 Artificial Neural Networks

This section deepens the knowledge of a particular machine learning model, i.e. ANNs. ANNs are function approximators and are considered in the remaining chapters only in the context of supervised learning. The following subsections give an overview starting from the general to the specifically used subtypes.

2.2.1 History

Artificial neurons are derived from the biological neurons of the human brain [Aiz11]. Therefore, it is not surprising that the first fundamental publications on this topic were published in the 1940s by psychologists such as Warren McCulloch [MP43] or Donald O. Hebb [Heb49]. If one compares the schematic illustration of a biological neuron (Fig. 2.2a) with today's representation of artificial neurons (Fig. 2.2b), various similarities can be observed. The biological neuron is connected by its dendrites (inputs) to axons (outputs) of other neurons. Synapses weigh the input signals at this junction before the signals are combined in the soma [Aiz11]. The resulting output signal is in turn transmitted via the axon. Despite the current analogies, the artificial neuron did not emerge in its present form due to one publication. While essential milestones in the history of ANNs are outlined below, please refer to [Sch14] for a more comprehensive review.

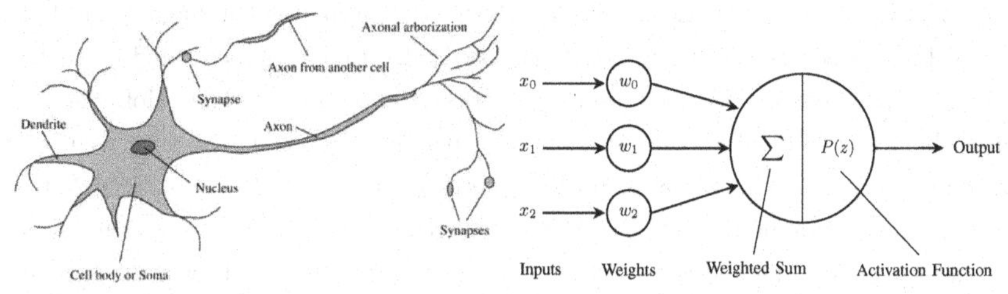

(a) Schematic of a biological neuron. With kind permission of the author taken from [RN10]

(b) Artificial Neuron based on [Gla21]

Figure 2.2: The Neuron

In 1943, McCulloch and Pitts mathematically described in [MP43] how neurons logically work. They remained limited to boolean values and utilized electronic logic gates. This ancient ANN could not learn. However, in 1949 Hebb provided the approach that between neurons firing simultaneously, the connections are strengthened [Heb49]. This fundamental principle is nowadays called *Hebbian learning*. In 1958 Rosenblatt further developed the simple neuron into the perceptron [Ros58]. He relaxed certain restrictions, e.g., hindrance of the weights to unity, introduced the bias as an additional parameter, and presented a simple but converging learning algorithm. This perceptron was limited to linear separable binary classifications but marked the first learning artificial neuron in history. In 1960, Widrow and Hoff delivered ADALINE [WH60], the first ANN that used the gradient descent algorithm. The error was no longer computed between the boolean class values but between the continuous weighted sum of the input values and the desired value. The mean squared error of the two quantities was then minimized by adjusting the weights using the derivative. Therefore, the ANN could learn for the first time even if there was no misclassification. In 1969, Minsky and Papert published [MP69], in which they pointed out the limitations of the basic perceptron, such as the inability to solve the simple XOR problem. Even though solutions to this problem already existed, the limited applicability led to a decline in funding [Cre93] and thus to the first so-called AI Winter, which lasted throughout the 1970s. Nevertheless, in 1971, Naum Aizenberg et al. presented in [AIP71] and [AIPK71] the multi-valued threshold function representing a neuron with complex numbers and allowing multiclass classification tasks. This was the first form of multi-valued neurons, which will be discussed in more detail in Section 2.2.3. However, since these contributions were written in Russian in the Soviet Union, they initially remained unknown to the research community. In 1974, Werbos adapted the backpropagation algorithm to ANNs in his dissertation [Wer74]. Rumelhart, Hinten, and Williams made it public in 1986 by showing in [RHW86] how it can be used to train multilayer neural networks by learning representative features. The period after that until the mid-1990s is referred to as the second AI Winter due to declining funding and lack of breakthroughs. As computing power increased, algorithms improved, and computations were parallelized, wider and deeper networks could be trained. Thus, Hofreiter and Schmidhuber revolutionized natural language processing (NLP) in 1997 with recurrent neural networks (RNN) and their Long Short-Term Memory (LSTM) method [HS97], the convolutional neural network (CNN) AlexNet [KSH12] marked a breakthrough in image recognition in 2012, and AlphaGo [SHM+16] was the first computer to win the game of Go against the world's best professional player at the time in 2016. Many improvements (e.g. dropout [SHK+14] or batch normalization [IS15]) have led to remarkable neural network performance. Still, these successes are mainly in the realm of narrow AI used on a problem-specific basis. Despite recent approaches like [BHX+22], humankind is still a long way from the promises of the early days regarding a general AI that can apply its knowledge in a wide variety of contexts or a super AI that exceeds human intelligence.

2.2.2 Architecture and Training

Let's now turn to the mathematical description of the artificial neuron, the structure of neural networks, and how neurons interact with each other. For this purpose, first, a particular type of architecture is examined. It is then detailed how information passes through the neural network and how it adapts itself based on the data.

Fully Connected Feedforward Network

Neurons can be interconnected to form neural networks to solve sophisticated tasks. They can be arranged in layers parallel to each other or stacked in several layers to increase their functionality [Aiz11]. The most basic form of such networks is the fully connected feedforward network, considered exclusively in the following chapters. Here, the output of each neuron is solely connected to the input of all neurons in the subsequent layer, i.e., there are no further connections to neurons in other layers [Gla21]. In the network, a distinction is made between the input, first hidden, hidden, and output layer, which are discussed in more detail in Chapter 2.2.4. The input layer does not contain any neurons but only the inputs. Figure 2.3 shows an example of a fully connected feedforward network architecture.

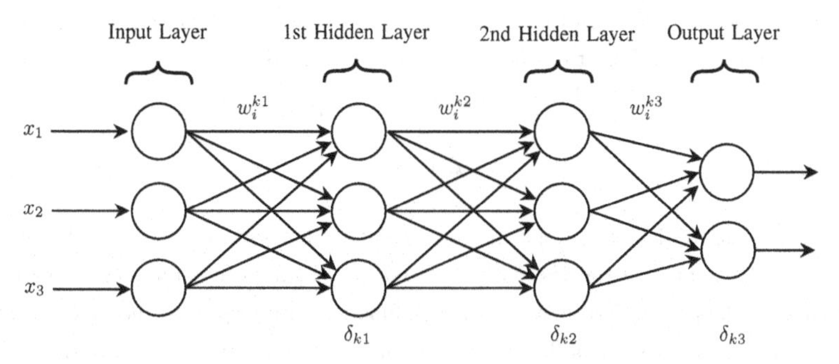

Figure 2.3: Fully connected feedforward network based on [Gla21] with three inputs x_i, three neurons in the first hidden layer, three neurons in the second hidden layer and two output neurons. Each connection between the layers is weighted with w_i^{kj}, where j denotes the corresponding layer, k the neuron, and i the connection. Only the bias is not shown due to space constraints. The individual error of the neuron is symbolized with δ_{kj}.

Forward Pass

As one can already guess from Figure 2.2b, the output y of a single neuron is calculated by

$$y = P(w_0x_0 + w_1x_1 + \cdots + w_nx_n), \tag{2.5}$$

where $x_0 = 1$, w_0 is the bias, w_i is the corresponding weight to the input x_i with $i = \{1, \ldots, n\}$ and $P()$ denotes the activation function. The activation function $P()$ can vary depending on the task of the neuron. By expanding from a single neuron to a layer, we write the calculations in matrix form, and by connecting m layers in series, we get the equation

$$Y_m = P_m(\ldots P_2(P_1(XW_1)W_2)\ldots W_m), \tag{2.6}$$

where Y_m denotes the output matrix, P_j is the activation function of the j-th layer, X is the input matrix and W_j is the weight matrix of the j-th layer. X is a matrix with dimensions *number of samples* x *number of features per sample + 1* and W_j is a matrix with dimensions *number of inputs + 1* x *number of neurons in the layer.* The additive one is due to the bias of the neurons. In classification tasks, the output layer assigns Y_m to the respective classes. Thus, the input values X are propagated forward through the network to predict the class label.

Backward Pass

To enable the neural network to learn from the data, it must evaluate its errors. Therefore, an error is mostly computed in the output layer, which is then propagated backwards through the network to adjust the weights. A standard method for this is backpropagation in combination with gradient descent. For this, a loss function $\mathcal{L}$ is defined that depends on the predicted value Y and the desired value D

$$\mathcal{L}(Y, D). \tag{2.7}$$

The more accurate the predictions are, the smaller the loss function becomes. Therefore, the goal is to minimize the loss function. Backpropagation accomplishes its part by determining the partial derivatives of the loss function with respect to the weights. This results in an optimization problem. To determine a local minimum resp. the global minimum for convex functions, the weights are adjusted using gradient descent by

$$\tilde{w} = w - \eta \cdot \frac{\partial \mathcal{L}(Y, D)}{\partial w}, \tag{2.8}$$

where η is an adjustable learning rate to regulate convergence to the minimum, w denotes the current weight, and $\tilde{w}$ thus represents the updated weight. The right term is subtracted as moving opposite the gradient, i.e., in the descent direction. There are many variants of the gradient descent, but almost all of them have the restriction that the loss function has to be differentiable resp. section-wise differentiable. Due to the chained structure of (2.6), the derivation is performed by the chain rule, which means that the activation functions must also be differentiable resp. section-wise differentiable. Another drawback is that the derivative per se has to be determined or at least estimated. The next chapter will deal with an alternative form of error backpropagation.

2.2.3 Multi-Valued Neuron

The mathematics of the following two sections is taken from [Aiz11], which is, among others, a summary of numerous publications of the author Igor Aizenberg and his father, Naum Aizenberg. Therefore, for simplicity, references are made only to what is beyond this work in these two sections. For basic derivations and convergence proofs, the reader is referred to [Aiz11]. Only the principles of the MVNs are presented here.

The MVN is a modification of a classical neuron. It has complex-valued inputs, weights, bias, output, and a non-differentiable activation function. The activation function maps depending on the weighted sum z to the unit circle, which is divided into k sectors described by the set

$$E_k = \{1, \varepsilon_k, \varepsilon_k^2, \dots, \varepsilon_k^{k-1}\}, \tag{2.9}$$

with $\varepsilon_k = e^{j\frac{2\pi}{k}}$, where j is the imaginary unit and $k \in \mathbb{N}_{>1}$. Therefore, the activation function of the continuous MVN is defined by

$$P(w_0x_0 + w_1x_1 + \dots + w_nx_n) = P(z) = e^{j\varphi} = \frac{z}{|z|}, \tag{2.10}$$

where w_0 is the bias, $x_0 = 1$, w_i is the corresponding weight to the input x_i with $i = \{1, \dots, n\}$ and $\varphi \in [0, 2\pi[$ is the argument of the weighted sum z. Fig. 2.4 illustrates this context. The discrete activation function differs only in that the phase is adjusted to the nearest bisector, i.e. $P(z) \in E_k \cdot e^{j\frac{\pi}{k}}$, where $e^{j\frac{\pi}{k}}$ realizes a shift of half a sector to move from the sector borders to the bisectors.

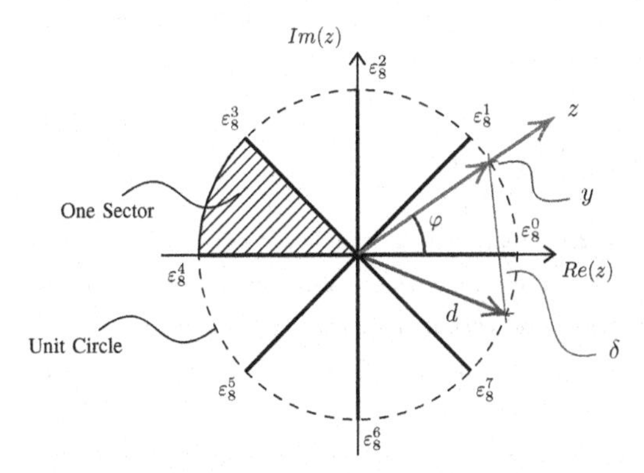

Figure 2.4: Diagram of a k-valued MVN based on [PL21]. The complex domain is reduced to the unit circle, which in turn is divided into $k = 8$ sectors. The weighted sum z with angle φ is mapped back to the unit circle by the activation function $P(z)$ and represents the current output of the neuron y. The desired output d is the bisector within the desired sector. The error δ defines the correction to get from y to d.

The MVN has a derivative-free learning rule determined by its error or, more precisely, by the correction to get from the current output to the desired one

$$\delta = d - z. \tag{2.11}$$

With the restriction that the inputs are located on the unit circle, the weights are updated through

$$\tilde{w}_i = w_i + \frac{\eta}{n+1} \cdot \bar{x}_i \cdot \delta. \tag{2.12}$$

To understand why this holds true, (2.11) is rearranged to d. This simultaneously yields the desired updated sum $\tilde{z}$

$$d = z + \delta \stackrel{!}{=} \tilde{z} = \tilde{w}_0 x_0 + \tilde{w}_1 x_1 + \cdots + \tilde{w}_n x_n. \tag{2.13}$$

By substituting equation (2.12) and with $\eta = 1$, it follows

$$\begin{aligned} \tilde{z} &= \tilde{w}_0 x_0 + \tilde{w}_1 x_1 + \cdots + \tilde{w}_n x_n \\ &= \left(w_0 + \frac{1}{n+1}\delta \bar{x}_0\right) x_0 + \left(w_1 + \frac{1}{n+1}\delta \bar{x}_1\right) x_1 + \cdots + \left(w_n + \frac{1}{n+1}\delta \bar{x}_n\right) x_n. \end{aligned} \tag{2.14}$$

Considering $x_i \bar{x}_i = 1$, since the inputs must lie on the unit circle, it can be concluded that

$$\begin{aligned} \tilde{z} &= w_0 x_0 + w_1 x_1 + \cdots + w_n x_n + \frac{1}{n+1}\delta x_1 \bar{x}_1 + \cdots + \frac{1}{n+1}\delta x_n \bar{x}_n \\ &= z + \frac{n+1}{n+1}\delta = z + \delta. \end{aligned} \tag{2.15}$$

Therefore, equation (2.12) represents a suitable way to adjust the weights. Nevertheless, a modified learning rule is used, which does not imply any structural changes to the update of the weights. However, for learning nonlinear input-output mappings, it is beneficial to compute the error using the projection of z onto the unit circle rather than the weighted sum z:

$$\delta = d - y. \tag{2.16}$$

Besides the derivative-free learning algorithm, the increased functionality of the MVN compared to a real-valued neuron is a significant advantage. The functionality can be enhanced by providing the perceptron with more neurons in both width and depth.

2.2.4 Multilayer Neural Network Based on Multi-Valued Neurons

The multi-layered feedforward structure with fully connected MVN is referred to as MLMVN. Considering an MLMVN structure $n - N_1 - \cdots - N_{m-1} - N_m$ with n inputs in the input layer, $m-1$ hidden layers, and the output layer m, the general learning case is taken into account. For this purpose, the algorithm is divided into three steps. Before starting the iterative algorithm, the weights are randomly initialized, and the biases are set to zero.

(i) The input x is forwarded through the network; hence the outputs are determined according to the learning sample and the current weights and biases.

(ii) Calculating the error of each neuron differs from a single MVN in that it is iterated over the output neurons and then the error is shared through the network. Therefore, the error δ is calculated beginning in the output layer similar to (2.16)

$$\delta_{km} = \frac{1}{N_{m-1}+1}(D_{km} - Y_{km}), \tag{2.17}$$

where k is the k-th neuron of the m-th layer, i.e., the last layer. Since a uniformly distributed error is assumed, it is equally divided by N_{m-1} the number of neurons from the preceding layer plus one for the bias. All further errors are computed backwards in the hidden layers by

$$\delta_{kj} = \frac{1}{N_{j-1}+1} \sum_{i=1}^{N_{j+1}} \delta_{i,j+1}(w_k^{i,j+1})^{-1}, \tag{2.18}$$

where j indicates the j-th layer, $N_0 = 0$ and $(w_k^{i,j+1})^{-1}$ is the inverse element of $w_k^{i,j+1}$, which is defined by

$$w^{-1} = \frac{\bar{w}}{|w|^2} \tag{2.19}$$

for complex numbers.

(iii) For weight adjustment three distictions are made: the first hidden layer, hidden layer 2 to $m-1$, and the output layer. Thereby the weights are updated successively from layer 1 to layer m.

1*st* hidden layer:

$$\tilde{W}_1 = W_1 + \frac{\eta}{(n+1)\cdot|z_{k,1}|}\bar{X}E_1, \tag{2.20}$$

hidden layer $2, \dots, m-1$

$$\tilde{W_j} = W_j + \frac{\eta}{(N_{j-1}+1) \cdot |z_{k,j}|} \bar{\tilde{Y}}_{j-1} E_j, \tag{2.21}$$

output layer m:

$$\tilde{W_m} = W_m + \frac{\eta}{(N_{m-1}+1)} \bar{\tilde{Y}}_{m-1} E_m, \tag{2.22}$$

where $\bar{\tilde{Y}}_{j-1}$ is matrix containing the updated complex conjugated output of the neurons from the $(j-1)$-th layer and E_j is the matrix holding the errors $\delta_{k,j}$ of the j-th layer. $\frac{1}{|z_{k,j}|}$ is an additional variable learning rate for nonlinear mappings that makes learning smoother. Here, the calculated error δ of a neuron is divided by the magnitude of its weighted sum z. The variable learning rate can be omitted in the output layer since the exact error is known here, and it is not computed heuristically as in the previous layers.

2.3 Loss Functions

Loss functions are key components of neural networks as they significantly influence the adjustment of the weights. Moreover, the selection of the loss functions depends on the task. For example, in regression, a common choice is the L_p norm [GBC16] of the error

$$\|x\|_p = \left(\sum_i |x_i|^p\right)^{\frac{1}{p}} \tag{2.23}$$

If one additionally divides the sum by the number of summands, the L_2 norm of the error with $p = 2$ is called root-mean-square-error (RMSE). However, multiclass cross-entropy, also called negative log likelihood [Bis06], is predominant in multinomial classifications. Here the loss of a training sample is calculated according to

$$\mathcal{L}(y, d) = -\sum_{k=0}^{K-1} d_k \cdot ln(y_k), \tag{2.24}$$

where k denotes the k-th class, d_k is a boolean variable indicating whether the classification is correct or not, and y_k is the normalized predicted probability of class membership. Chapter 3 will clarify this further. Since MLMVN are used, and no probabilities are output in the classical sense, the loss function $\mathcal{L}$ is already predetermined by equation (2.16). However, as the function can have complex values and $\mathbb{C}$ does not represent an ordered field compared to $\mathbb{R}$, an additional parameter is introduced that quantifies the loss. A predetermined threshold of this loss may be used to terminate the learning process since datasets cannot always be learned thoroughly. The error is calculated as the angular difference [Aiz11] of the desired output and the actual output

$$\Gamma \equiv arg(Y_m) - arg(D_m) \pmod{2\pi}, \quad D_m \in E_k \cdot e^{j\frac{\pi}{k}} \tag{2.25}$$

and is mapped on the interval $[0, 2\pi[$. The mean-square-error (MSE) is determined according to

$$MSE = \frac{1}{N_m} \sum_{s=1}^{N_m} \gamma_s^2, \tag{2.26}$$

where γ_s are the values of the N_m sized colum vector Γ.

2.4 Summary

This chapter has presented the necessary preliminaries for this thesis. Basic concepts of ML have been explained and put into context in the sense that one is dealing with supervised learning in the domain of multinomial classifications. Specific performance metrics were introduced to allow for comparable results. The emergence of MVNs was elaborated in the context of the history of ANNs. Their architecture was limited to the fully connected feedforward network. It was found that MVNs are complex-valued neurons that do not require derivations in their learning algorithm compared to BP combined with some form of GD. An additional loss function based on the angular difference was introduced to quantify the loss. The next chapter addresses the state of the art in neural network classification tasks and compares real-valued artificial neural networks (RVNN) with complex-valued artificial neural networks (CVNN).

Chapter 3

Scientific State of the Art

Neural networks are benchmark classifiers in many areas, especially for classification tasks [DLLT21]. However, depending on the particular type of classification data, ANNs differ in terms of architecture, training methods, and data preprocessing. Convolutional networks, for example, are especially suited for processing a grid of values, such as an image [GBC16]. Recurrent networks achieve state memory through their recurrent connections and are thus explicitly suited for processing sequences of values [GBC16]. Even combinations of these architectures can lead to outstanding results [BHX$^+$22]. However, for the sake of not going beyond the scope, the following state of the art analysis will be limited to typical building blocks of classification, such as the activation functions. Furthermore, special attention will be paid to the output layer and the assignment of neurons to individual classes. After briefly introducing multinomial classification, the chapter splits the topic into using real-valued (Chapter 3.1) and complex-valued neural networks (Chapter 3.2). Whereby MLMVN is dealt with explicitly. Chapter 3.3 complements this by discussing the presented state of the art methods. Finally, Chapter 3.4 summarizes the most important findings.

Multinomial classification, also referred to as multi-class classification, is a task that uses supervised learning. As mentioned in Chapter 2.1, sample x is assigned to a target value y. This is distinguished from multi-label classification, which can assign multiple target values y_i to a sample x. The target value can be represented as an integer value $y \in \{0, 1, \ldots, k-1\}$, where k is the number of classes and no longer contains categorical classifications in text form. Furthermore, the target value can be one-hot encoded, resulting in a binary representation of the target value, in the sense that only the digits 1 and 0 are used for class indication. The target class is assigned the value 1 and all other classes are set to 0. Considering the example of a 6-class problem with $y \in \{0, 1, 2, 3, 4, 5\}$, $y_i = 2$ corresponds to the one-hot encoded vector $[0, 0, 1, 0, 0, 0]$. Each digit corresponds to exactly one of the six classes.

J. Knaup, *Impact of Class Assignment on Multinomial Classification Using Multi-Valued Neurons*, BestMasters, https://doi.org/10.1007/978-3-658-38955-0_3

3.1 Multinomial Classification in Real-Valued Neural Networks

Real-valued neural networks achieve benchmark results [BMR+20][DLLT21] on well-known datasets such as ImageNet [DDS+09] in computer vision or Penn Treebank [MMS93] in natural language processing. To achieve these results, there are several possible settings. One of these possibilities is the selection of a suitable activation function. Activation functions should be nonlinear to enable the learning of nonlinearly separable datasets [GBC16]. An overview of many real-valued activation functions with their application area is given in [Sza21]. This section addresses three well-known activation functions (Fig. 3.1). The S-shaped sigmoid function is bounded $\sigma : \mathbb{R} \rightarrow]0,1[$

$$\sigma(x) = \frac{e^x}{e^x+1} = \frac{1}{1+e^{-x}}. \tag{3.1}$$

In contrast to the ancestral threshold function, i.e. the step function, the sigmoid function is differentiable and thus can be used in combination with backpropagation. However, due to its boundedness to the range interval]0,1[, which may be advantageous for probability theory, it is prone to the vanishing gradient problem [BSF94]. Furthermore, it is not 0 centred, which triggers instabilities in the learning behaviour [Sza21]. At least this point, the hyperbolic tangent remedies by $\tanh : \mathbb{R} \rightarrow]-1,1[$. It can be described as a scaled sigmoid function

$$2\sigma(2x) - 1 = tanh(x) = \frac{sinh(x)}{cosh(x)} = \frac{e^x - e^{-x}}{e^x + e^{-x}}. \tag{3.2}$$

Both functions are suitable for binary classification in the output layer due to their two-sided boundedness. Therefore, they can be applied both in a 2-valued classification problem or for a multi-label classification in which each output neuron represents a digit in the binary output vector. However, a different activation function has been established for the hidden units [Sza21], namely the Rectified Linear Unit (ReLU) [NH10]

$$ReLU(x) = max(0,x). \tag{3.3}$$

The ReLU is also not 0 centred, but it counteracts the vanishing gradient problem in the sense that it has a constant derivative for values $x > 0$. The derivative of the sigmoid functions decreases by approaching their limit values. However, ReLU also has drawbacks, such as the dying ReLU problem [Sza21], where too many neurons in the network are disabled and do not provide gradients. Therefore, variations such as leaky ReLU are gaining popularity [Gla21].

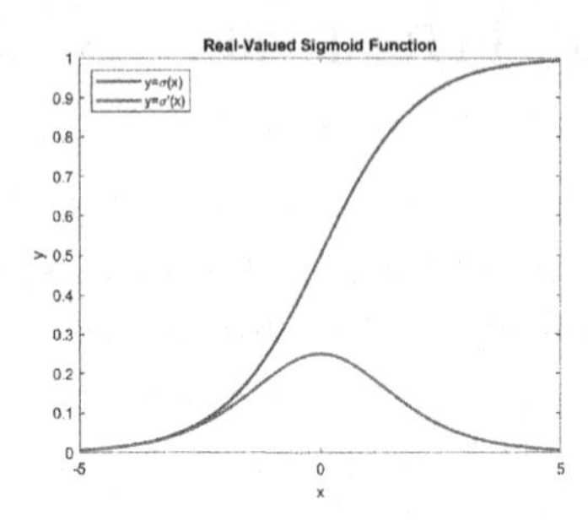

(a) Real-valued sigmoid function with its derivative.

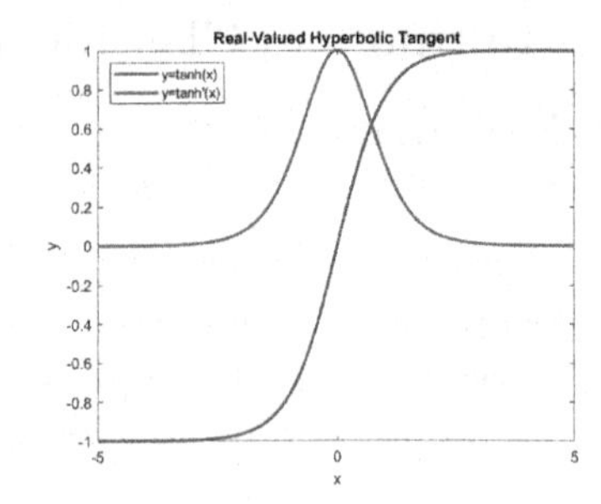

(b) Real-valued hyperbolic tangent with its derivative.

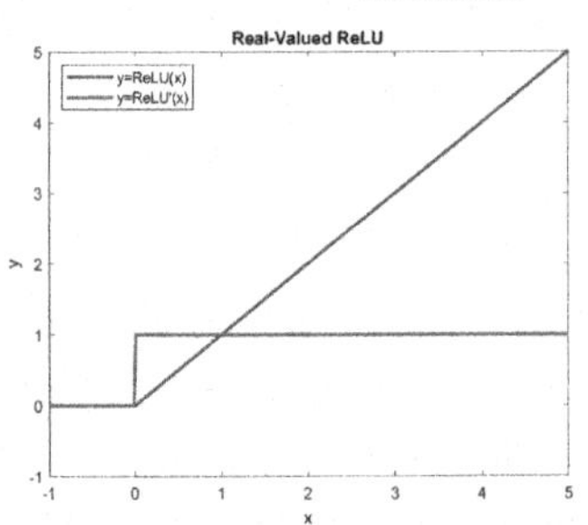

(c) Real-valued ReLU with its derivative. Since the derivative does not exist at point $x = 0$, it was arbitrarily set to 1. The smooth line is due to the representation, the derivative is not continuous in $x = 0$.

Figure 3.1: Real-valued activation functions with their derivatives.

Another fundamental building block for multinomial classification is the softmax function

$$softmax(x_k) = \frac{e^{x_k}}{\sum_{k=0}^{K-1} e^{x_k}}, \tag{3.4}$$

or normalized exponential, which can be considered as a multi-class generalization of the sigmoid function [Bis06]. Previous activation functions have been only applied to single neurons. However, for a multinomial classification ($k > 2$), it is necessary to include all possible classes and calculate the respective probabilities for an assignment. Therefore, there are as many neurons in the output layer as classes. The softmax function adds up the values of all output neurons and divides the individual values by the sum. This results in values for each class in the interval $]0, 1[$. These single values add up to 1 and are interpreted as the probability of class membership. Furthermore, they can be evaluated with the loss function presented in chapter 2, i.e. the negative log-likelihood.

3.2 Multinomial Classification in Complex-Valued Neural Networks

CVNNs differ from RVNNs by having complex-valued inputs, weights, and activation functions. Due to their increased capacity, they can outperform RVNNs, which are comparable in their number of real-valued parameters [BRM$^+$21]. Nevertheless, it was shown in [MM18] that CVNNs perform worse when they receive real-valued inputs or when no meaningful transformation of the input data could take place. In this case, the imaginary part often followed the real part, leading to increased computational effort and no added value. As complex numbers extend real-valued numbers by one imaginary dimension, complex numbers can be expanded by further imaginary dimensions to hypercomplex numbers. However, hypercomplex ANNs are not considered here, although the mathematics presented below can be extended to them. The use of, e.g. quaternions ($\mathbb{H}$), also has certain advantages but is too small a niche [BQL21] to be carried out here.

In general, to train CVNNs using BP, the activation functions must be differentiable in the complex domain. Complex differentiation is a stronger property than in the real-valued case. Here, not only must the limit

$$f'(z_0) = \lim_{z \to z_0} \frac{f(z) - f(z_0)}{z - z_0} \tag{3.5}$$

exist for a function $f : \mathbb{C} \to \mathbb{C}$ to be differentiable at a point z_0 on an open subset $\Omega \subset \mathbb{C}$ [MM18]. A complex function must also satisfy the Cauchy-Riemann equation (3.6) [Cau25]. Thereby a complex function $f(z)$ is decomposed into the real differentiable functions $\Re(f(z)) = u(x, y)$ and $\Im(f(z)) = v(x, y)$ by $f(z) = u + jv$ [BSMM15] with $z = x + jy$ and $z \in \mathbb{C}, x, y \in \mathbb{R}$. If the limit (3.5) exists and the partial derivatives satisfy

$$\frac{\partial u}{\partial x} = \frac{\partial v}{\partial y}, \qquad -\frac{\partial u}{\partial y} = \frac{\partial v}{\partial x}, \tag{3.6}$$

a differentiable function f is also referred to as holomorphic, entire, or analytic, at least in the univariate case. The often-quoted Liouville theorem that a holomorphic bounded function $f : \mathbb{C} \to \mathbb{C}$ must be constant results in an activation function being unbounded and/or non-holomorphic [BP92][Hir12][MM18]. The theorem does not consider the multivariant case $f : \mathbb{C}^n \to \mathbb{C}^n$, but there is hardly any literature on this, so it remains a point of further investigation. Nevertheless, many non-holomorphic functions are not entirely differentiable but can be evaluated in restricted regions. To calculate the derivatives of holomorphic and non-holomorphic [AAANM11] functions efficiently, the Wirtinger calculus [Wir27] is used

$$\frac{\partial}{\partial z} = \frac{1}{2}\left(\frac{\partial}{\partial x} - j\frac{\partial}{\partial y}\right), \qquad \frac{\partial}{\partial \bar{z}} = \frac{1}{2}\left(\frac{\partial}{\partial x} + j\frac{\partial}{\partial y}\right). \tag{3.7}$$

Here the function $f(z)$ is taken as a function of z and $\bar{z}$, which avoids a separate inspection of the real and imaginary part, where for holomorphic functions $\frac{\partial}{\partial \bar{z}} = 0$ holds [Fis02]. As a consequence, the chain rule for complex multivariate functions [MM18] is given by

$$\begin{aligned} \frac{\partial}{\partial z_i}(f \circ g) &= \sum_{k=1}^{n}\left(\frac{\partial f}{\partial z_k} \circ g\right)\frac{\partial g_k}{\partial z_i} + \sum_{k=1}^{n}\left(\frac{\partial f}{\partial \bar{z}_k} \circ g\right)\frac{\partial \bar{g}_k}{\partial z_i} \\ \frac{\partial}{\partial \bar{z}_i}(f \circ g) &= \sum_{k=1}^{n}\left(\frac{\partial f}{\partial z_k} \circ g\right)\frac{\partial g_k}{\partial \bar{z}_i} + \sum_{k=1}^{n}\left(\frac{\partial f}{\partial \bar{z}_k} \circ g\right)\frac{\partial \bar{g}_k}{\partial \bar{z}_i} \end{aligned}. \tag{3.8}$$

This mathematical background allows the evaluation of our activation functions $P(z)$. Activation functions can be applied to z in the complex domain [MM18] or the single real-valued parts $\Re(z)$ and $\Im(z)$ in the Cartesian representation [Hir12], respectively magnitude of z and phase of z in polar representation [SVVHU18]. Activation functions applied separately to the real and imaginary parts are referred to in the literature as type A, and activation functions related to the polar representation are designated as type B. The activation function $P(z)$ with the hyperbolic tangent is therefore expressed in the three variations (Fig. 3.2-3.4)

$$P_1(z) = tanh(z) = \frac{sinh(z)}{cosh(z)} = \frac{e^z - e^{-z}}{e^z + e^{-z}}, \tag{3.9a}$$

$$P_2(z) = tanh(\Re(z)) + j \cdot tanh(\Im(z)), \tag{3.9b}$$

$$P_3(z) = tanh(|z|) \cdot e^{j \cdot arg(z)}, \tag{3.9c}$$

where (3.9a) represents an unbounded function. For the complex ReLU, however, this distinction is not possible. Since the $max()$ operator does not work in the unordered field $\mathbb{C}$ in the classical sense and a $max()$ operation does not influence the magnitude of z, only the application to the real and imaginary part remains here

$$ReLU(z) = ReLU(\Re(z)) + j \cdot ReLU(\Im(z)) = max(0, \Re(z)) + j \cdot max(0, \Im(z)). \tag{3.10}$$

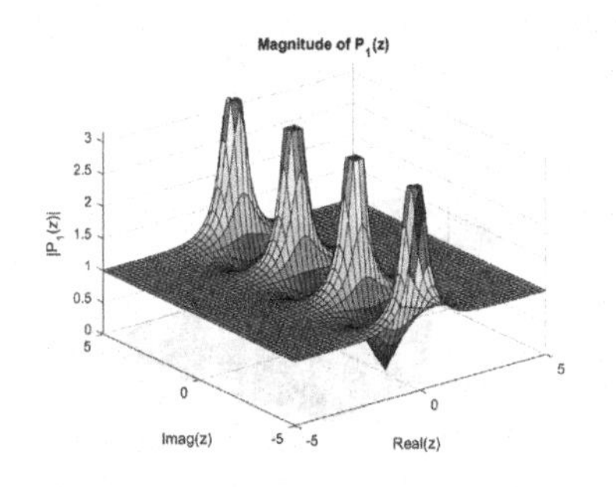

(a) Complex surface plot of $|P_1(z)|$ with singularities for $z = \frac{j}{2} \pm kj\pi, k \in \mathbb{N}_0$

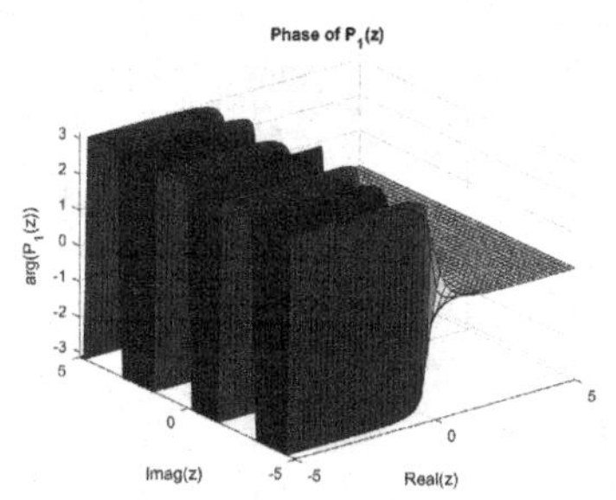

(b) Complex surface plot of $arg(P_1(z))$

Figure 3.2: Complex-valued hyperbolic tangent

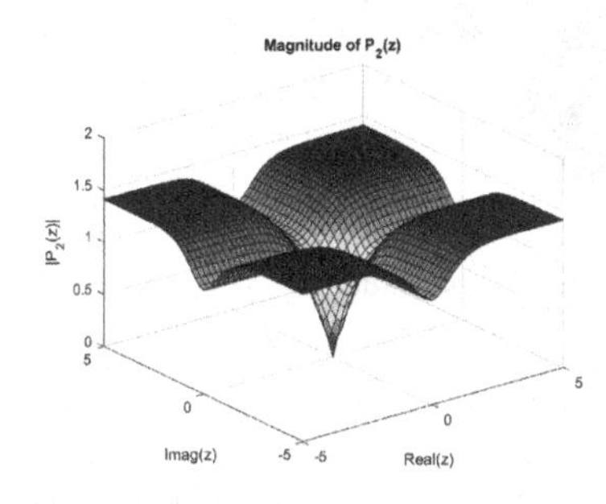

(a) Complex surface plot of $|P_2(z)|$

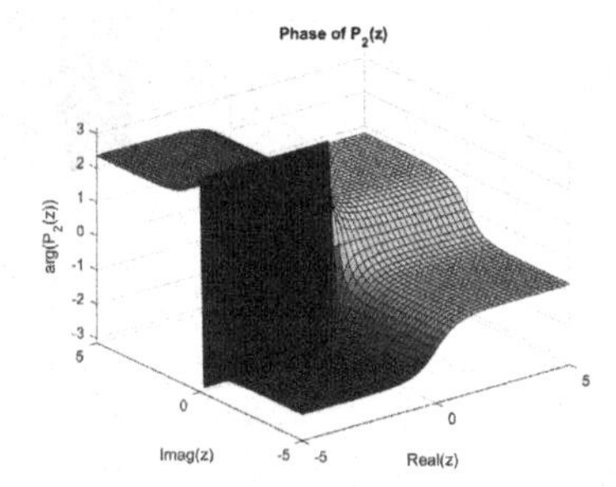

(b) Complex surface plot of $arg(P_2(z))$

Figure 3.3: Hyperbolic tangent applied to $\Re(z)$ and $\Im(z)$

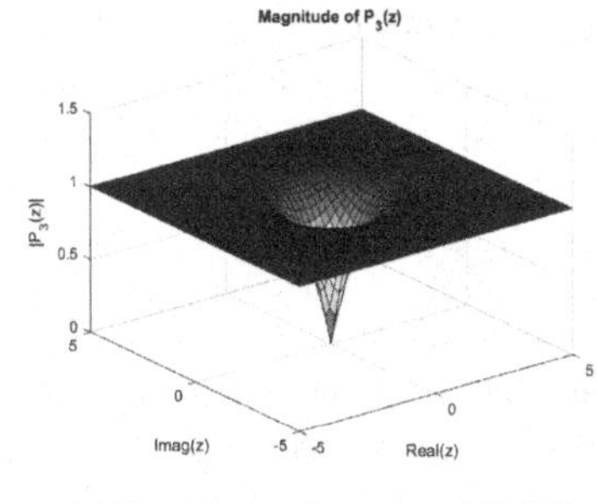

(a) Complex surface plot of $|P_3(z)|$

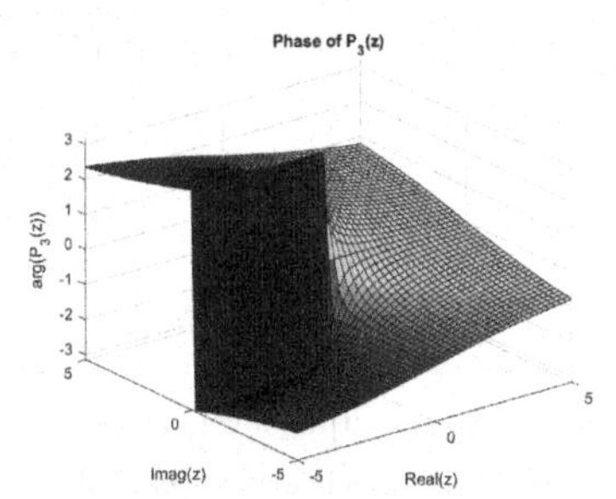

(b) Complex surface plot of $arg(P_3(z))$

Figure 3.4: Hyperbolic tangent applied to $|z|$

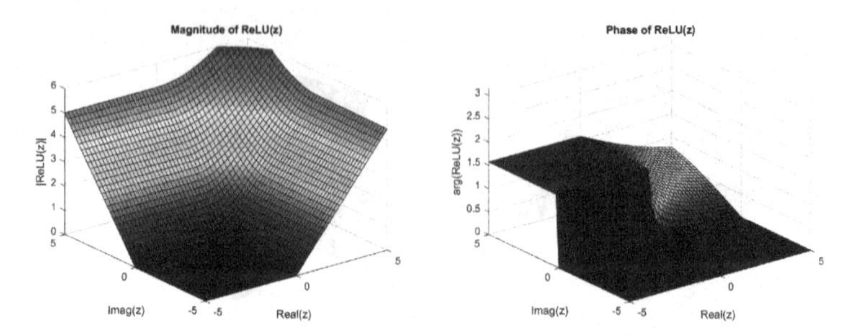

(a) Complex surface plot of $|ReLU(z)|$ (b) Complex surface plot of $arg(ReLU(z))$

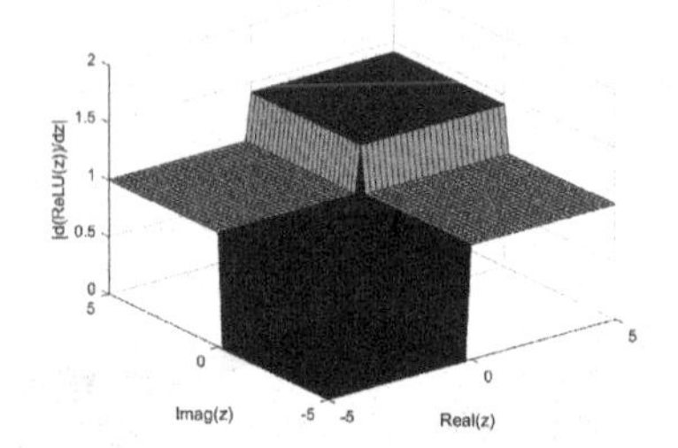

(c) Complex surface plot of the magnitude of the derivative of $ReLU(z)$

Figure 3.5: $ReLU(z)$

As with RVNNs, CVNNs have also shown that, in general, the ReLU function often achieves better results than other hidden unit activation functions [MM18][BQL21]. In this context, other modifications of the complex ReLU exist, but [TBZ+18] have shown that the ReLU by definition from (3.10) outperforms it. In Figure 3.5c, it can be seen that the derivative of the complex ReLU weights one quadrant stronger than the others.

$$\sigma(|z|^2) = \frac{e^{x^2+y^2}}{e^{x^2+y^2}+1} = \frac{1}{1+e^{-x^2-y^2}}. \tag{3.11}$$

$$softmax(|z_k|^2) = \frac{e^{x_k^2+y_k^2}}{\sum_{k=0}^{K-1} e^{x_k^2+y_k^2}} \tag{3.12}$$

For evaluating the output layer, the sigmoid (for multi-label) and the softmax function (for multinomial classifications) can also be used in CVNNs. There are also as many neurons in the output layer as classes. However, the complex inputs must first be mapped into $\mathbb{R}$ using squared magnitudes to achieve a real-valued loss. Thus, functions (3.1) and (3.4) are used in $\mathbb{R}$ change to functions (3.11) and (3.12) used in the complex domain.

MLMVN

As introduced in Section 2.2.3, MVNs have the particular activation function (2.10) that projects z onto the unit circle. Thus, an evaluation by sigmoid or softmax is not suitable either. Hence, the loss function (2.26) presented in Chapter 2.3 was introduced. There is a single output neuron for smaller datasets with a small number of classes [Aiz11], such as the iris dataset [Fis36]. By dividing the unit circle into k or a periodic multiple of k sectors (Fig. 3.6a), the neuron can assign the appropriate class to the input value depending on the angle of z. For larger datasets such as the MNIST dataset for handwritten digits [LC10] with ten classes, MLMVN also use as many output neurons as there are classes [AG18]. Here, a neuron divides the unit circle into two sectors (Fig. 3.6b). A value in the upper half means that this neuron does not symbolize the desired class. Whereas a value in the bottom half indicates that the neuron represents the corresponding class. However, this can result in error cases where a) several neurons indicate to be the correct class and b) no neuron indicates a class membership. For the error case a) the *winner takes it all* technique was introduced in [AG18]. Accordingly, the neuron with the smallest error with respect to the desired bisector is selected. In [AABF00], test set samples with error case b) were moved into the learning set.

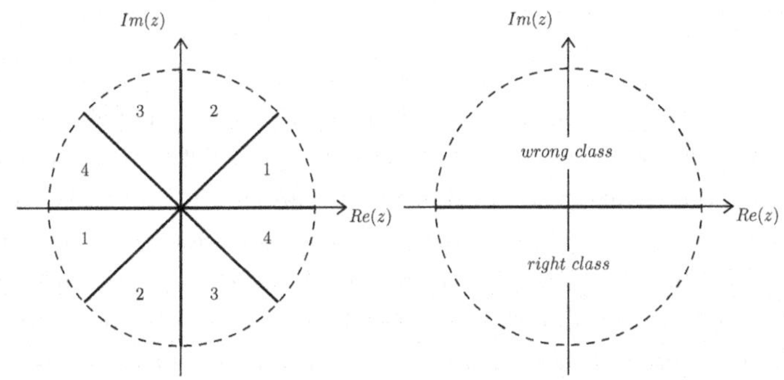

(a) Diagram of k-valued output neuron. The unit circle is divided into $k \cdot p = 8$ sectors with $k = 4$ and $p = 2$, where k denotes the number of classes and p the periodicity.

(b) Diagram of one of k output neurons. The unit circle is divided into two sectors. Each neuron symbolizes a class and indicates a class member in a boolean way. If z is in the upper half, the input value belongs to another class; in the lower half, class membership is indicated.

Figure 3.6: Multinomial Classification with MVN in the output layer

3.3 Discussion

As we have seen before, both RVNNs and CVNNs achieve benchmark results in classification tasks [DLLT21][TBZ+18]. State of the art networks use both BP to update their weights. RVNNs use well-known activation functions for this purpose. For CVNNs, the choice of activation function is a more contentious area. Although it has recently been shown that the complex ReLU gives reasonably strong results similar to its real-world counterpart [TBZ+18], the activation function selection is still a recent research topic. CVNNs have found less application than RVNNs in the past. This is due to several reasons. First, complex arithmetic operations have an increased computational cost depending on the representation form (polar or cartesian) [MM18]. Second, they do not perform better than RVNNs on real-valued data that cannot be transformed meaningfully to the complex domain [MM18]. However, due to the extension of essential training methods, such as batch normalization and weight initialization, to complex numbers [TBZ+18], implementing CVNNs is becoming more straightforward. Since CVNNs are also less prone to overfitting than RVNNs [BRM+21], interest in them is growing. However, it is more challenging to train complex networks since stronger properties are required to derive complex-valued functions. But by using holomorphic functions, due to the Cauchy-Riemann equation, it is sufficient to compute only two instead of four gradients.

Nonetheless, MLMVN circumvent these difficulties by using a derivative-free algorithm to update their complex weights [Aiz11]. Therefore, the non-differentiable but nonlinear activation function (2.10) can be used. However, this discards the magnitude component, allowing the complex number to contain less information than with conventional activation functions. Nevertheless, MVNs offer the possibility to replace multiple output neurons. Publications show both the case where only one output neuron is used [Aiz11] and the case where MLMVNs, like RVNNs and other CVNNs, use as many output neurons as classes [AG18]. However, there is no literature for multiple output neurons with a multiple threshold logic. Chapter 4 provides an approach to this specific problem. In particular, it is necessary to clarify how classes are allocated to neurons and how they should be arranged on the unit circle.

3.4 Summary

This chapter gave a state of the art overview of essential components of multinomial classification using ANNs. It was stated that activation functions must be nonlinear to learn nonlinearly separable datasets [GBC16]. Furthermore, differences between real-valued and complex-valued activation functions were presented. In particular, properties were treated that a function must satisfy to be differentiable in the complex domain. It was further shown that MLMVN could reduce the number of output neurons compared to RVNNs and conventional CVNNs. The next chapter will show whether this is possible without performance loss and how this is achieved.

Chapter 4

Approach

This chapter presents a novel approach that utilizes multiple non-binary MVNs in an output layer. For this purpose, the ML chain is defined so that both an image dataset and a classical dataset with numerical features can be evaluated in the following. In chapter 4.1, the data preparation and processing are discussed, on which a feature engineering is carried out in chapter 4.2. In these two chapters, the focus is not on maximizing classification accuracy but on evaluating existing methods and establishing a comparable baseline. The approach to exploit multiple non-binary MVNs is explained in chapter 4.3. The evaluation model is presented in chapter 4.4 before chapter 4.5 provides a summary.

4.1 Data Preparation and Processing

Before insights can be gained from data, the data must be put into an appropriate form so that the algorithm can process it. The implementation described in Chapter 5.1 requires data that is both numeric and has no missing values. The selected datasets from Chapter 5.2 already fulfil these requirements; hence no further preparations are necessary. Furthermore, the datasets contain additional preprocessing steps, which are referred to in the corresponding chapter. Moreover, the input values must be located on the unit circle in the complex plane, as mentioned in Chapter 2.2.3, so that the learning algorithm of the MLMVN can be exploited. This transformation can be accomplished in several ways. Therefore, applied methods to perform the transformation are presented below for both use cases.

J. Knaup, *Impact of Class Assignment on Multinomial Classification Using Multi-Valued Neurons*, BestMasters, https://doi.org/10.1007/978-3-658-38955-0_4

Numeric Dataset

The first method is suitable for real-valued datasets. Each input value x_i is transformed to the unit circle in the complex plane using $e^{j\varphi_i}$. The calculation of φ_i is obtained by the min-max scaler [Aiz11]

$$\varphi_i = \frac{x_i - a}{b - a} \cdot \alpha, \tag{4.1}$$

where a is the minimum of the real-valued feature and b is the respective maximum. The factor $\alpha \in [0, 2\pi[$ is applied for the scaling on the unit circle. Since the circularity of the phase at $\alpha \approx 2\pi$ causes the min and max values to be close to one another on the unit circle, α may need to be adjusted depending on the application. Outliers should be removed in advance; otherwise, the linear phase scaling will be shifted.

Image Dataset

The second method can be applied to images, although the following is limited to grayscale images. Since representative features shall be learned to perform image classification and Oppenheim has shown in [OL81] that the phase component of the Fourier transform contains most of the information about edges and their spatial position, the images are transformed into the frequency domain. Thereby the Fourier transform represents one of many possibilities to perform a transformation into the frequency domain. Here, the two dimensional discrete Fourier transform (2D DFT) will be employed since it has already found application with MLMVNs in both [AABF00] and [AG18]. In [AG18] it was shown that MLMVNs with this preprocessing achieved 100% accuracy on the MNIST dataset for handwritten digits [LC10]. Even though it was corrected in an errata that this only referred to the learning set and 90% accuracy was achieved on the test set, this approach will be re-evaluated here.

An MxN dimensional image has the real-valued function values $f[m, n]$. The 2D DFT is given by [Bov05]

$$F[k, l] = \frac{1}{MN} \sum_{m=0}^{M-1} \sum_{n=0}^{N-1} f[m, n] \cdot e^{-j2\pi\left(\frac{km}{M} + \frac{ln}{N}\right)}. \tag{4.2}$$

Only the phase information is taken into account to get values on the unit circle, and the magnitude is set to unity. On the one hand, MVNs have the advantage here that they treat the circularity of the phase appropriately [Aiz11]. In contrast to the min-max scaler (4.1), it is desired that phase values like $2\pi - 0.01$ and $2\pi + 0.01$ are close to each other [AG18]. On the other hand, the 2D DFT loses properties by discarding the magnitude. Thus, for instance, only the amplitude spectrum is invariant to translation [Bov05]

$$f[m - m_0, n - n_0] \longleftrightarrow F[k, l] \cdot e^{-j2\pi\left(\frac{m_0 m}{M} + \frac{n_0 n}{N}\right)}. \tag{4.3}$$

Therefore, as in the MNIST dataset for handwritten digits, centred images are required.

4.2 Feature Engineering

Feature engineering is a key component of ML to create meaningful input values. The selection of particular features additionally controls the dimensionality of the data to ensure computability within a reasonable time. For example, the Fisher discriminant ratio (FDR) [Fis36] can evaluate significant features. Techniques, such as combinatorial refinement [DL14], achieve a dimensionality reduction through the weighted combination of individual features. Despite these tools, feature engineering can also model expert knowledge. Thus, feature engineering might be context-dependent as well as dataset-dependent.

Here, processing and feature engineering become blurred since the previous chapter explained the transformation to the unit circle in the complex plane. In the case of the numerical dataset, normalization of the data has been done by (4.1). Due to the already reduced dimensionality of the following dataset; no further feature engineering is done here. In the case of the image dataset, the dimensionality of the input variable of MxN is present with the 2D DFT. To reduce the number of features, only a subset of the pixels of the 2D Fourier spectrum is selected. In [AR75], same classes were found to have similar coefficients in the low-frequency domain. Furthermore, [AG18] stated that in an MxM image, the $\lfloor M/m \rfloor$-th frequency is required to detect a structure of size mxm. This led to selecting the lowest three frequencies for the MNIST dataset of handwritten digits. Therefore, only low-frequency components of the phase spectrum are considered in the following.

In the MxN image of the Fourier spectrum, low frequencies are located in the upper left corner (see Fig. 4.1a). However, the image can be periodically extended [BB08]. Therefore, a shift of all pixels by $\frac{M}{2}$ or $\frac{N}{2}$ in the respective direction leads to the centring

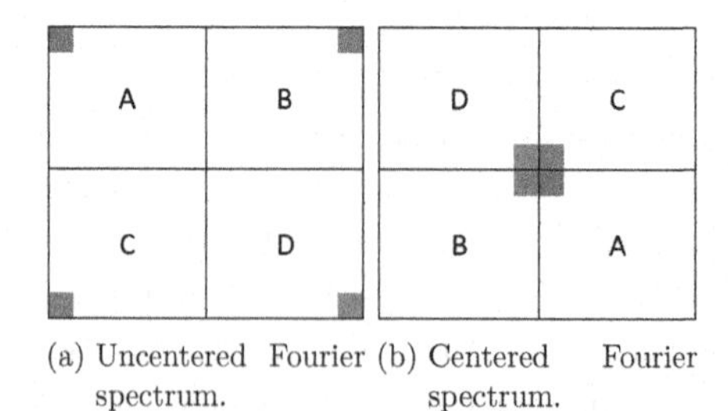

(a) Uncentered Fourier spectrum. (b) Centered Fourier spectrum.

Figure 4.1: Centering of the 2D DFT spectrum by swapping the image quadrants. Based on [BB08]. (a) Low frequency coefficients are located at the origin in the upper left corner and due to the periodicity also in the other corners. (b) Low frequency coefficients are located in the center of the image.

of the low-frequency components [BB08]. The shift corresponds to the swapping of the quadrants, according to Fig. 4.1. The different frequencies subsequently lie on radial trajectories and can be selected with the help of the masks shown in Fig. 4.2 [Bov05]. The desired pixel set P of the low-frequency image components is described by the inside of a circle. In [AG18] Aizenberg approximates these circles with the *diamond rule* using squares to represent the rings. Here, however, the pixel selection is described through the circle equation by

$$P = \left\{ F[k,l] \middle| \left(k - \frac{M}{2}\right)^2 + \left(l - \frac{N}{2}\right)^2 \leq r^2 \right\}, \tag{4.4}$$

where r indicates the circle's radius. The only exception is the initial pixel $F[0,0]$ respectively, now the shifted centre pixel. This can be interpreted as the mean intensity of the image, but for real-valued images, it is also real-valued and contains no phase information. Therefore [AG18] has discarded this pixel. However, due to its meaning, the min-max scaler (4.1) is applied to this pixel here. By choosing $r = 4$, a set of features with 49 pixels is obtained, which on the one hand, is sufficiently large to learn from and, on the other hand, does not result in matrix sizes that are too extensive. As with the numerical dataset, the data is passed to the ANN as a feature vector, i.e., the two-dimensional image becomes a one-dimensional row vector.

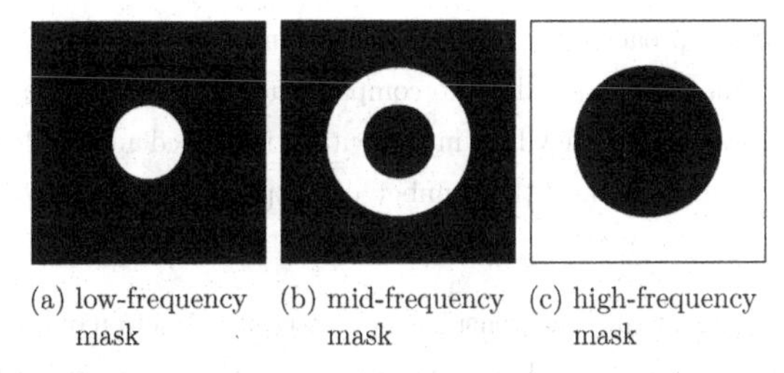

(a) low-frequency mask (b) mid-frequency mask (c) high-frequency mask

Figure 4.2: Masks for radial frequency selection. Black part is multiplied by 0 and white part is multiplied by 1. Based on [Bov05].

4.3 Classification

This section explains the approach of assigning classes to multiple non-binary MVNs in the output layer to perform classifications. The terms *assignment*, *allocation*, and *arrangement* have to be defined first in this context. Assignment is declared as an umbrella term and includes both allocation and arrangement. Allocation refers to the mapping of which neuron represents which classes, and arrangement refers to the placement of the class sectors on the unit circle. Therefore the section is divided into two subsections, Class Allocation and Class Arrangement.

4.3.1 Class Allocation

This subsection is devoted to the question of which classes should be mapped to which output neuron. It has already been described in Chapter 3 that, to the best of the author's knowledge, there is no literature on this topic at the current stage. Furthermore, current implementations leverage MVNs only as binary classifiers [AG18] or as single output neurons [Aiz11]. In the following, different approaches to class allocation will be presented.

First, both existing approaches, binary assignment and classification by a single output neuron, will be evaluated. This is done to compare the later results. On the other hand, the number of classes per neuron will be incrementally increased and randomly distributed among the output neurons, i.e., the number of output neurons will be incrementally reduced.

Another way of looking at this is to consider the classes as a function of their features. It is to investigate whether it is reasonable to allocate similar classes in the feature space to the same neuron. In analogy to the Hebbian learning [Heb49] is this based on the assumption that such a mapping strengthens the connections between the relevant features and the corresponding output neurons. This potentially allows the weighted sum to be more accurately determined than if it has to differentiate between dissimilar classes. However, assessing the similarity of classes is not trivial, and it should not be assumed that an ANN automatically classifies based on the same decision criteria as a human. In [Yud08], it is described that an ANN should detect camouflaged tanks in front of a forest. For the test and training data, this worked with 100% accuracy. However, in another test set, it was noticed that the ANN learned to distinguish between sunny and cloudy days. In fact, the photos with tanks were taken on a cloudy day, whereas the photos of the forest were taken on a sunny day. Another example is provided by [RSG16], where huskies and wolves were not distinguished based on their external appearance, but on the criterion of whether there was snow in the background. Therefore, four methods are employed to determine the similarity of the classes based on the features. The four selected methods allow dimension reduction to represent similarities graphically. Besides, they are widespread methods, whereby they are already available in function libraries. Since they are only a means to an end, they are only briefly presented here.

- PCA: The first method is the principal component analysis (PCA) [Pea01]. PCA is an unsupervised method that performs a projection into the lower dimensional space, the principal components [Alp14]. Here, the first principal component is the projection, where the variance is the largest [Alp14]. The second principal component is also said to maximize the variance but is orthogonal to the first principal component [Alp14].
- LDA: The second method is linear discriminant analysis (LDA) [Fis36]. LDA is a supervised method and seeks a projection that keeps the data as separable as possible, in the sense that the class means are far apart and the class variance is as small as possible [Alp14].
- t-SNE: The third method is t-distributed stochastic neighbor embedding (t-SNE) [vdMH08]. t-SNE converts euclidean distances from high dimensional space into conditional probabilities and reconstructs a two or three-dimensional space to represent the data points [vdMH08].
- UMAP: The fourth and last method is Uniform Manifold Approximation and Projection for Dimension Reduction (UMAP) [MHM18]. UMAP is mainly based on manifold theory and topological data analysis, where it can represent the global data structure more accurately than t-SNE [MHM18]. UMAP and t-SNE are unsupervised methods.

After assessing the similarity of the classes, the classes are allocated to the output neurons. The number of output neurons is also reduced incrementally, i.e. the number of classes per neuron is increased so that a comparison with the randomly assigned classes is possible. Here, both variants are tested, mapping similar classes to a neuron and distinct classes. Thus, the hypothesis is to test whether the assignment of similar classes positively affects the network connections and hence the metrics.

Another point worth mentioning is the extension of one-hot encoding from Chapter 3. By allocating multiple classes to a neuron, it can index more than just 0s and 1s. Recalling the 6-class example with $y \in \{0, 1, 2, 3, 4, 5\}$, and allocating three classes to one neuron, each of the two neurons has four sectors. Three sectors indicate the respective class, and one additional sector denotes that the class does not belong to this neuron. This reduces the one-hot encoded vector corresponding to $y_i = 2$ to $[3, 0]$. Since all neurons within a layer are treated the same way, a particular error case may occur. If the number of classes is not divisible by the number of neurons, sectors are created in the last neuron, which cannot be assigned to any class. Thus classes can be predicted, which do not exist in the dataset. However, since there are no training samples for the respective class, a mapping in the later stages of training in this sector should be relatively rare.

4.3.2 Class Arrangement

This subsection is devoted to the question of how to arrange the class sectors on the unit circle. The periodic extension of the activation function (see Fig. 3.1a) is not considered. The periodicity is specifically helpful for learning non-linearly separable data for single neurons, as in solving the XOR problem with one neuron [Aiz11]. It can also benefit multicluster classifications [Aiz11]; however, this is not the focus here.

As shown in (4.9), the weight adjustment Δw_i is proportional to the error δ. The magnitude of the error always lies in the interval $[0, 2]$. The minimum error $\delta = 0$ is present when $P(z)$ is equal to the desired bisector d. The maximum error $\delta = 2$ is present when $P(z)$ is opposite on the unit circle to the desired bisector d. Larger values are not possible due to the activation function (2.10).

As with the class allocation from Chapter 4.3.1, there is no literature on this subject. Therefore, a reasonable hypothesis is also proposed, which has to be verified in Chapter 5. Since similar classes are defined to have similar feature values, it is more difficult for the weighted sum z to take on very different values for similar classes. It is more intuitive that different feature values will result in different weighted sums for distinct classes. Therefore, it is assumed that it makes more sense to arrange similar classes next to each other on the unit circle rather than opposite each other.

Since the error for continuous neurons can take all values in the interval $[0, 2]$, the mathematical description starts from the average error of each class. This means that for this purpose, we consider the discrete neuron, which additionally maps $P(z)$ to the nearest bisector, i. e., $P(z) \in E_k \cdot e^{j\frac{\pi}{k}}$ (see (2.9) and the following paragraph). For an MVN with k sectors, n indicates how many sectors $P(z)$ and the desired bisector d_k are apart. Suppose they are identical then $n = 0$, for adjacent sectors $n = 1$ and so on. The direction of the adjacency does not matter. Since $P(z)$ and d_k always form an isosceles triangle with two

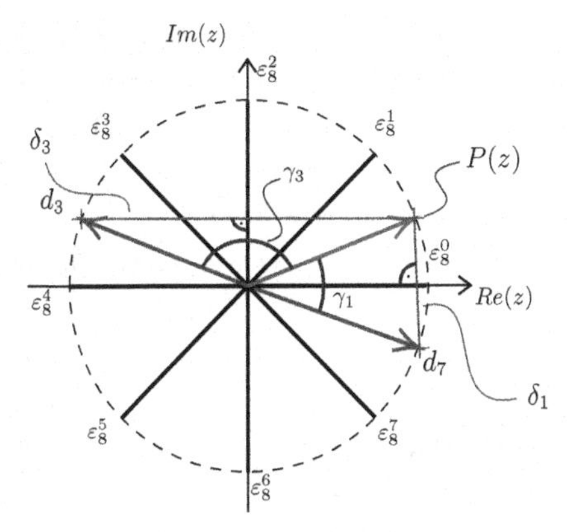

Figure 4.3: Diagram of a k-valued discrete MVN based on [PL21]. The complex domain is reduced to the unit circle, which in turn is divided into $k = 8$ sectors. The weighted sum z is mapped back to the unit circle by the activation function $P(z)$ and represents the current output of the neuron. The desired outputs d_k are the bisectors within the desired sectors. $P(z)$ is mapped to the nearest bisector for the discrete neuron. The error δ_1 is given if the desired bisector d_k and $P(z)$ are one sector apart. The direction of the separation, clockwise or counterclockwise, is irrelevant. If the desired bisector d_k and $P(z)$ are three sectors apart, the error is δ_3. The minimum error $|\delta_0| = 0$ is present when $P(z)$ is on the desired bisector d_k. If the desired bisector lies between ε_8^4 and ε_8^5, the maximum error $|\delta_4| = 2$ would be obtained. For discrete neurons, $P(z)$ and the desired bisector d_k form an isosceles triangle with the two sides of length 1.

side lengths of 1, the third side length δ_n can be calculated by

$$|\delta_{disc,n}| = 2 \cdot |P(z)| \cdot \sin\left(\frac{\gamma_n}{2}\right) = 2 \cdot \sin\left(\frac{\gamma_n}{2}\right). \tag{4.5}$$

The angle γ_n is given by

$$\gamma_n = \frac{2\pi}{k} \cdot n, \tag{4.6}$$

which finally results in the error magnitude

$$|\delta_{disc,n}| = 2 \cdot \sin\left(\frac{\pi}{k} \cdot n\right). \tag{4.7}$$

The more classes are assigned to a neuron, the smaller the error between adjacent classes becomes. Oppositely arranged classes keep a large error value. This supports the hypothesis that similar classes should be adjacent. Indeed, larger errors result in more extensive weight adjustments, allowing the weighted sum z to be shifted more quickly.

The arrangement of neurons on the unit circle depends on the dataset and is therefore carried out in Chapter 5.2. The order of the specified class numbers of a neuron determines the arrangement. Figure 4.3 serves as an example with seven classes per neuron. Bisector d_0, i.e. the bisector between ε_8^0 and ε_8^1, is reserved for all classes that do not belong to this neuron. Thus, neurons have at least one sector more than they can assign to classes. If there is only one output neuron, this rule does not apply. Afterwards, the classes are assigned counterclockwise to the sectors in the listed order. The class number listed last corresponds to d_7.

4.4 Model Training

This subsection introduces the ML model used for training on the numerical and image dataset. For this purpose, an architecture is defined, and corresponding hyperparameters are specified. Moreover, a modification for the loss is introduced. Furthermore, initializations and employed training methods are presented, such as a batch algorithm.

Since MVNs are utilized as output neurons, the ML model is an MLMVN. The error calculation (2.16) is applied for a more accurate mapping of the nonlinear input-output connections [Aiz11]. In terms of architecture, it has been shown, in contrast to conventional ANNs, that shallow networks with a single hidden layer often perform best [PL21][AG18]. Therefore, only one hidden layer and one output layer are employed here as well. For the numerical dataset presented in the next chapter, it has been shown that one hidden layer with 100 neurons performs best [PL21]. For the image dataset, the evaluation of the MNIST dataset is considered. There, one hidden layer was utilized with 1024 neurons and increased successively in powers of 2 to improve the accuracy [AG18]. Since class assignment is to be investigated and the focus is not on maximizing the accuracy of a particular dataset, the basic number of neurons is chosen. This results in 100 hidden neurons for the numerical dataset and 1024 hidden neurons for the image dataset. The number of output neurons is varied and determined depending on the assessment in Chapter 5.2. The learning rate is set to $\eta = 1$ and the scaling factor of the min-max scaler is fixed to $\alpha = 1.99\pi$ to approximately exploit the entire unit circle.

In contrast to [PL21] and [AG18], no soft margin technique is applied. The soft margin technique [Aiz14] freezes weights at an error threshold to converge the learning behaviour. Here, on the other hand, weights are adjusted as long as there is an error. To achieve the best result in the end, the test set is evaluated after each training epoch. If the updated weights exceed the best result, these weights are stored. The stored weights are loaded for the final evaluation instead of the last training weights. This prevents the learning from being interrupted due to an artifical threshold value.

The datasets are randomly shuffled if they are not already shuffled. A seed value is given to the random number generator so that the same random shuffle is present for each test. Unless the datasets have their own training/test split, 80% of the data are assigned to the training set and 20% to the test set. Again, the distribution is implemented with a pseudorandom mechanism resulting in the same split for all test runs. The number of training epochs is set to 100.

Two special error cases can occur when dealing with multiple MVNs in the output layer, as mentioned in Chapter 3.2. In case a), various neurons indicate that they symbolize the class by multiple nonzero entries in the output vector. In case b), no neuron indicates class membership, i. e., each entry of the output vector is 0. In case a), the described *winner takes it all* technique is applied. Thus, the neuron is selected, which has the smallest error to the next bisector with respect to the angle. In case b), the *winner takes it all* rule is extended such that the neuron farthest from the bisector of 0 is mapped to the closest adjacent bisector.

Due to the subtraction and the modulo operator, the loss function (2.25) decreases as y approaches d in the clockwise direction. The change from the minimum (0) to the maximum value ($\approx 2\pi$) occurs when y exceeds d in the clockwise direction. The loss function becomes independent of the direction of the angle by computing it with this modification:

$$\Gamma = \pi - ||arg(Y) - arg(D)| - \pi|. \tag{4.8}$$

The modified loss function decreases for the convergence of y to d from both directions. In contrast to (2.25), the loss function (4.8) maps to the interval $[0, \pi[$.

Batch Algorithm

To speed up learning, multiple samples are processed at once. This subset from the learning set is called a batch, which gives another hyperparameter, the batch size. The algorithm presented in [ALM12] introduces the notation

$$\Delta w_i = \frac{1}{n+1}\delta\bar{x}_i \tag{4.9}$$

for this purpose. Revisiting equation (2.15), it follows

$$\delta = \Delta w_0 x_0 + \Delta w_1 x_1 + \cdots + \Delta w_n x_n. \tag{4.10}$$

Considering a batch size of S samples, a system of linear equations is obtained

$$\begin{aligned}
\Delta w_0 x_0^1 + \Delta w_1 x_1^1 + \dots + \Delta w_n x_n^1 &= \delta^1 \\
\Delta w_0 x_0^2 + \Delta w_1 x_1^2 + \dots + \Delta w_n x_n^2 &= \delta^2 \\
\vdots \qquad\quad \vdots \qquad\qquad\quad \vdots \quad &\quad \vdots \\
\Delta w_0 x_0^S + \Delta w_1 x_1^S + \dots + \Delta w_n x_n^S &= \delta^S
\end{aligned} \tag{4.11}$$

which can be extended to multiple neurons [AA14] and written in generalized form

$$X \Delta W = E. \tag{4.12}$$

A solution for ΔW can be computed in different ways. Therefore, several cases of consideration arise.

(i) If X is a square matrix with $det(X) \neq 0$, both sides can be multiplied from the left with X^{-1} to get a unique solution for ΔW.

(ii) If $n > S$, it is an underdetermined system, and in general, there are infinitely many possible solutions.

(iii) If $n < S$, it is an overdetermined system, and there may be no solution.

In cases (ii) and (iii), the Moore-Penrose pseudoinverse can be determined by first computing a singular value decomposition (SVD) [Hog07] of the complex matrix X

$$X = U \Sigma V^H, \tag{4.13}$$

where U and V are unitary matrices and therefore $V^H = V^{-1}$. The superscript H indicates the conjugate transpose operator. Σ is a real-valued diagonal matrix. The pseudoinverse of X is then given by

$$X^{\dagger} = V \Sigma^{\dagger} U^H, \tag{4.14}$$

with

$$\Sigma^{\dagger} = \begin{cases} \frac{1}{\sigma_{ij}} & \text{if } i = j \land \sigma_{ij} \neq 0 \\ 0 & \text{otherwise} \end{cases}. \tag{4.15}$$

An approximation to ΔW can be obtained by substituting equation (4.13) in (4.12) and multiplying by the pseudoinverse $X^{\dagger}$ from the left.

$$\begin{aligned}
U \Sigma V^H \Delta W &= E \\
V \Sigma^{\dagger} U^H U \Sigma V^H \Delta W &= V \Sigma^{\dagger} U^H E \\
\Delta \breve{W} &= X^{\dagger} E
\end{aligned} \tag{4.16}$$

For case (ii), $X^\dagger$ is the pseudoinverse with minimum L_2 norm $\|\Delta\breve{W}\|_2$, and for case (iii), $X^\dagger$ minimizes the L_2 norm of $\|X\Delta\breve{W} - E\|_2$ [GBC16].

Since former implementations did not support subdivision of training data into batches, the entire training set is often taken as a batch [PL21]. To reduce matrix sizes, [AG18] takes a batch size of 500. Hence a batch size of approximately 500 is specified. The batch size should be chosen to be a divisor of the number of training samples so that all batches have the same size. If this is not feasible, the training and test split is slightly changed to allow the training data to be divisible by a number of approximately 500.

Initialization

A well-chosen starting point in the weight space can be beneficial for fast convergence of the learning algorithm. In this context, initialization for CVNNs is still a subject of current research. In [TBZ+18], a sophisticated method for deep CVNNs is presented. However, since shallow CVNNs are trained here, standard initializations are adopted.

In general, small numbers are used to avoid exploding and vanishing gradients [BSF94]. However, since no derivatives are involved, this does not apply here. Nevertheless, the output values should remain close to the unit circle to keep the errors according to (2.11) small. Therefore, initializations are used that are symmetrically distributed around 0 and use small numbers. The random numbers drawn from the given distributions are applied to both the real and imaginary parts of the weights. All biases are set to 0.

The first distribution

$$W \sim \mathcal{U}(-0.5, 0.5) \tag{4.17}$$

is based on [Aiz11] and its MLMVN implementation in Matlab. However, it is not explicitly designed for MVNs and their activation function. The second type

$$W \sim \mathcal{U}\left(-\frac{1}{\sqrt{m}}, \frac{1}{\sqrt{m}}\right) \tag{4.18}$$

represents a common initialization and is a widely used standard interval, where m is the number of inputs to the layer [GBC16]. In [GB10], Glorot and Bengio show that the interval

$$W \sim \mathcal{U}\left(-\sqrt{\frac{6}{m+n}}, \sqrt{\frac{6}{m+n}}\right) \tag{4.19}$$

is a faster converging alternative for deep neural networks, where n is an additional parameter representing the number of outputs in the layer. This initialization is referred to as normalized Xavier initialization. The Kaiming initialization

$$W \sim \mathcal{N}\left(0, \frac{2}{m}\right), \tag{4.20}$$

or He initialization [HZRS15], is specific to deep ANNs with the nonlinear ReLU activation function. It is based on a normal distribution, unlike the other approaches. However, preliminary tests with the chosen architectures did not lead to noticeable differences. Therefore, only the initialization (4.18) is chosen in the following to keep the number of tests low.

4.5 Summary

In this chapter, the approach for appropriate class assignment on multiple MVNs was presented. For this purpose, the necessary preprocessing was outlined first. Furthermore, the min-max scaler was described as a possibility to transform numerical data to the unit circle. For image datasets, the phase information of the 2D DFT was utilized to map the pixel values onto the unit circle. Especially low-frequency components are relevant for classification. It was hypothesized that similar classes in feature space should be allocated to the same neuron and arranged adjacent to each other. Subsequently, hyperparameters for the training model were established, a loss modification was introduced, a batch algorithm for MLMVNs was presented, and an initialization method was selected.

Chapter 5

Evaluation

This chapter evaluates the approach presented in Chapter 4. First, Chapter 5.1 outlines implementations details of the MLMVN model. Subsequently, Chapter 5.2 presents two datasets and examines them for their class similarity. Afterwards, Chapter 5.3 provides the results of the evaluation. Finally, Chapter 5.4 concludes with a discussion and Chapter 5.5 summarizes the evaluation.

5.1 Implementation

The model described in Chapter 4.4 is implemented in the open-source framework PyTorch [PGM+19]. The model was created in analogy to torch.nn.Module, which is the base class of all neural net classes in PyTorch. Thus, the object-oriented model can be stacked with any number of layers, and the number of neurons per layer can be chosen arbitrarily. Due to the implementation with tensors, the computations can be executed either on a CPU or a GPU. Even though the computations in this chapter were performed on a CPU, the implementation is the first to leverage MLMVNs on GPUs utilizing the described batch algorithm. The repository extends a pre-created baseline version of MLMVNs. The source code can be downloaded from the Nextcloud. The individual contributions to the repository can be viewed in the Git History at `https://ds-juist.init.th-owl.de/j.knaup/CVNNMVN`. For PCA, LDA, t-SNE, and the metrics, the library scikit-learn [PVG+11] was employed. Based on this, [MHSG18] was chosen for the implementation of UMAP.

J. Knaup, *Impact of Class Assignment on Multinomial Classification Using Multi-Valued Neurons*, BestMasters, https://doi.org/10.1007/978-3-658-38955-0_5

5.2 Datasets

This section introduces both datasets, the numerical one and the image dataset. The two datasets are examined for class similarity based on their features, and an assignment of the classes to the corresponding neurons is made for testing.

5.2.1 Sensorless Drive Diagnosis

The numerical Sensorless Drive Diagnosis (SDD) dataset [DG17] consists of features generated from two phase currents of a test motor placed in a modular demonstrator [BDML12]. The demonstrator is composed of a test motor, measuring shaft, bearing module, flywheel and load motor [ZLH+12]. Various fault conditions can be installed by synthetically corrupted hardware. The demonstrator was created in the German Federal Ministry of Economics and Technology research project, named Autonomous Drive Technology by Sensor Fusion for Intelligent, Simulation-based Production Facility Monitoring and Control (AutASS in German). Empirical Mode Decomposition [HSL+98] was used to determine three intrinsic mode functions and their residuals for each phase current [BDML12]. The statistical values mean, standard deviation, skewness and kurtosis were subsequently calculated for these feature vectors. These statistical parameters constitute the 48 features of the SDD dataset. For a more detailed description regarding the demonstrator setup and the feature description, please refer to [BERB+13], [BDML12] and [ZLH+12].

The dataset consists of 58509 samples, contains 11 different classes and is balanced. Here, class 1 symbolizes the fault-free state of the engine, and the other ten classes are fault cases. The fault cases can result from a shaft misalignment, an axis inclination or a bearing failure. Table 5.1 gives an overview of the classes and the respective fault cases. The 80/20 split between training and test set results in 46806 samples in the training set and 11703 samples in the test set. The training set is further divided into 87 batches of 538 samples each.

Table 5.1: Error indicators of the individual classes of the SDD dataset. Class 1 is error-free, and the remaining ten are error cases. Equal classes, such as class 4 and 5, are not identical; they differ in the level of error, for example, the angle of the axis inclination.

Class	1	2	3	4	5	6	7	8	9	10	11
Bearing Failure	0	0	0	0	0	1	1	1	1	1	1
Axis Inclination	0	0	1	1	1	0	1	1	0	1	1
Shaft Misalignment	0	1	0	1	1	1	0	1	1	0	1

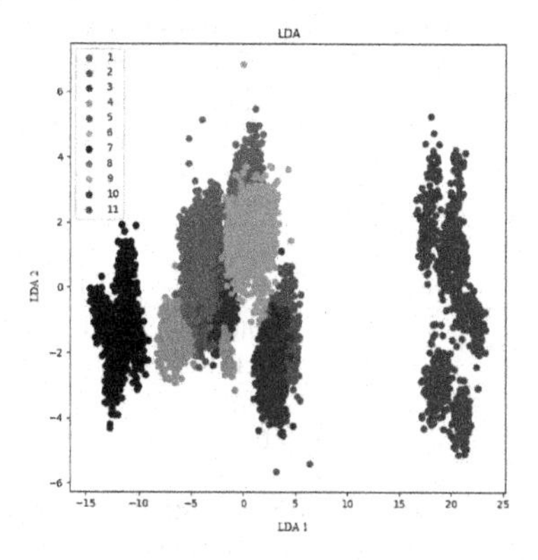

Figure 5.1: Feature space of the SDD dataset projected to 2D via LDA. Class 11 and 7 are linearly separable from the remaining classes. The further classes intersect each other.

For checking the similarities of the classes, the four methods presented in Chapter 4.3.1, PCA, LDA, t-SNE and UMAP, are applied. On this dataset, the supervised method LDA achieved the most decisive result. Even additional standardization of the data did not achieve improvements of the other methods over LDA. From Figure 5.1, it can be seen that class 11 is linearly separable from the rest of the classes. Also, class 7 does not intersect with other classes. Contrary to an intuitive assumption, the error-free class 1 is strongly overlapping and primarily similar to classes 6 and 9. Even though these specific findings are only valid for the selected feature space, it serves as the basis for evaluation. Therefore, test for class allocation are performed according to Table 5.2 and test for class arrangement are performed according to Table 5.3. Since only the model name will be referenced in the following, it will be briefly explained here. The capital letter symbolizes the corresponding dataset (A for AutASS) and the first index denotes the number of output neurons. The second index indicates whether the test is an allocation or an arrangement test. Finally, S symbolizes whether similar classes are adjacent resp. mapped to one neuron, or D symbolizes whether distinct classes are adjacent resp. mapped to one neuron.

Table 5.2: Allocation of the classes of the SDD dataset to the output neurons. The first model index shows the number of output neurons, and the second indicates an allocation test. For the third index, S denotes that similar classes were mapped to a neuron, and D denotes that distinct classes were mapped to a neuron. The number of classes per neuron is one less than the number of sectors per neuron.

Model	Cl/N	Neuron 1	Neuron 2	Neuron 3	Neuron 4	Neuron 5	Neuron 6
$A_{6,Al,S}$	2	1,7	2,10	3,4	5,8	6,9	11
$A_{6,Al,D}$	2	1,4	2,8	3,7	5,10	6,11	9
$A_{4,Al,S}$	3	1,6,9	2,3,10	4,5,8	7,11	-	-
$A_{4,Al,D}$	3	1,2,3	4,6,10	5,7,11	8,9	-	-
$A_{3,Al,S}$	4	2,7,10,11	3,4,5,8	1,6,9	-	-	-
$A_{3,Al,D}$	4	1,2,5,7	3,8,9,11	4,6,10	-	-	-

Table 5.3: Arrangement of the classes of the SDD dataset on the output neuron. The first model index shows the number of output neurons, and the second indicates an arrangement test. For the third index, S denotes that similar classes were mapped adjacent, and D denotes that distinct classes were mapped adjacent.

Model	Cl/N	Neuron 1
$A_{1,Arr,S}$	11	7,4,3,5,8,1,6,9,2,10,11
$A_{1,Arr,D}$	11	1,4,6,7,9,11,8,2,5,10,3

5.2.2 Fashion MNIST

The Fashion MNIST dataset [XRV17] consists of images of clothing items from Zalando. Since the MNIST dataset for handwritten digits [LC10] is deprecated, and classification algorithms achieve over 99% accuracy on it, the Fashion MNIST dataset is the more sophisticated successor [XRV17]. The datasets are identical in structure to test the same algorithms on the Fashion MNIST dataset as on the MNIST dataset for handwritten digits with as little effort as possible. Therefore, the fashion MNIST dataset has also ten different classes and grayscale images in 28x28 format. Figure 5.2 shows an example image from each class. The dataset contains 70000 samples, of which 60000 form the training set and 10000 form the test set. The samples are randomly shuffled. Each class is represented by 7000 samples, resulting in a balanced dataset. The grayscale images are 8-bit coded, giving intensity values from 0 to 255, and the image intensities are negated, causing the background to be black. Further preprocessing steps can be taken from [XRV17].

When evaluating the class similarities of the Fashion MNIST dataset, t-SNE and UMAP provide the most interpretable results. Figure 5.3 shows the impact of the two methods on the original data. The evaluation of the preprocessed features does not offer an interpretable result. As seen in both images, the classes can be divided into the subgroups of shoes, trousers, bags and tops. The shoe subgroup comprises the three classes sandals, sneakers and ankle boots. In the subgroups trousers and bags, there are only the classes

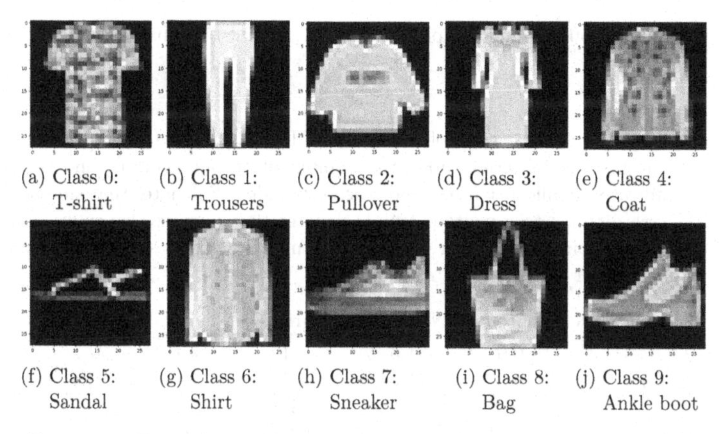

(a) Class 0: T-shirt (b) Class 1: Trousers (c) Class 2: Pullover (d) Class 3: Dress (e) Class 4: Coat

(f) Class 5: Sandal (g) Class 6: Shirt (h) Class 7: Sneaker (i) Class 8: Bag (j) Class 9: Ankle boot

Figure 5.2: Example images of all classes of the Fashion MNIST dataset

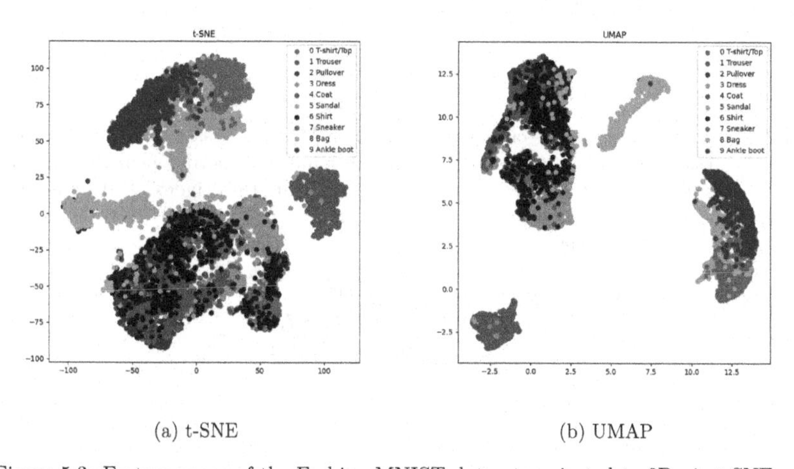

(a) t-SNE (b) UMAP

Figure 5.3: Feature space of the Fashion MNIST dataset projected to 2D via t-SNE and UMAP. The classes show subgroups of shoes, trousers, bags and tops. The shoes subset contains classes 5, 7, and 9, the pants subgroup consists of class 1 and the bags subgroup of class 8. The tops set is composed of classes 0, 2, 3, 4, and 6.

with identical names. T-shirts, pullovers, dresses, coats, and shirts form the tops subgroup. This categorization is intuitive and would probably have been made by a human being in the same way. Within the tops subgroup, it can be seen that dresses are on the outer edge. Pullovers and coats, in particular, overlap significantly. Shirts intersect primarily with t-shirts, pullovers, and coats. Based on these findings, the allocation test is performed according to Table 5.4, and the arrangement test is performed according to

Table 5.5. The 60000 training samples are subdivided into 120 batches of 500 samples each.

Table 5.4: Allocation of the classes of the Fashion MNIST dataset to the output neurons. The first model index shows the number of output neurons, and the second indicates an allocation test. For the third index, S denotes that similar classes were mapped to a neuron, and D denotes that distinct classes were mapped to a neuron. The number of classes per neuron is one less than the number of sectors per neuron.

Model	S/N	Neuron 1	Neuron 2	Neuron 3	Neuron 4	Neuron 5
$F_{5,Al,S}$	3	0,6	1,3	2,4	5,9	7,8
$F_{5,Al,D}$	3	0,9	1,6	2,7	3,5	4,8
$F_{4,Al,S}$	4	0,6,3	2,4,8	7,5,9	1	-
$F_{4,Al,D}$	4	5,2,3	7,6,1	9,0,8	4	-
$F_{3,Al,S}$	5	0,6,4,2	8,7,5,9	3,1	-	-
$F_{3,Al,D}$	5	1,3,4,5	0,2,7,8	6,9	-	-

Table 5.5: Arrangement of the classes of the Fashion MNIST dataset on the output neuron. The first model index shows the number of output neurons, and the second indicates an arrangement test. For the third index, S denotes that similar classes were mapped adjacent, and D denotes that distinct classes were mapped adjacent.

Model	S/N	Neuron 1
$F_{1,Arr,S}$	10	1,3,0,6,4,2,8,7,5,9
$F_{1,Arr,D}$	10	0,9,2,1,6,5,3,8,4,7

5.3 Results

After applying the methods from Chapter 4 to the datasets from Chapter 5.2, this section presents the test results. For this purpose, the section is divided into subsections for *class allocation* and *class arrangement* results. For the sake of clarity, only global metrics are presented in tabular form, as well as representative curve plots. For single class performances and further time history statistics, please refer to Appendix A for the majority of the data.

5.3.1 Class Allocation

Table 5.6 shows the results for the allocation tests on the Fashion MNIST and SDD datasets. It can be observed that models A_{11} and F_{10}, i.e., binary allocation of classes to neurons, performs best on the respective dataset. Compared to [PL21], the accuracy on the SDD dataset was improved from 88% to 96% with this implementation. In [XRV17], multilayer perceptrons have a classification rate of 84% to 87% on the Fashion MNIST dataset. The result falls minimally short of this value. Compared to the best-known result with 96.91% accuracy from [TKK20], the difference is more significant. When increasing the number of sectors per neuron from two to three, performance loss can be observed. However, there is hardly any difference in the global metrics between similar and distinct allocated classes. Nevertheless, from graphs A.5 and A.14, it can be noticed that the respective S models learn faster, resulting in a steeper increase in accuracy in the left part of the graphs. Increasing the sectors per neuron again, it becomes clear that allocations of similar classes to a neuron outperform allocations of distinct classes. However, it can be seen that the classification rate decreases with the increased number of sectors per neuron. Representing all training histories, it can be seen in Figure 5.4 and 5.5 that for the SDD dataset, the test accuracy tracks the training accuracy; for the Fashion MNIST case, the test accuracy stagnates early, and only the training accuracy continues to improve. However, it is not overfitting because the test accuracy does not decrease. In all cases, the training loss for the S models drops faster than for the D models.

Table 5.6: Test results for class allocation on the Fashion MNIST and AutaASS datasets. The superior result is shown in bold per comparison case. As the number of sectors per neuron increases, the metric values decrease. The test cases with similarly allocated classes almost exclusively outperform the distinct ones.

Model	S/N	Accuracy	Precision	Recall	F1-Score
A_{11}	2	**0.96**	**0.96**	**0.96**	**0.96**
$A_{6,Al,S}$	3	**0.90**	0.90	**0.90**	0.90
$A_{6,Al,D}$	3	**0.90**	**0.91**	**0.90**	**0.91**
$A_{4,Al,S}$	4	**0.71**	**0.76**	**0.71**	**0.72**
$A_{4,Al,D}$	4	0.64	0.71	0.64	0.66
$A_{3,Al,S}$	5	**0.66**	**0.69**	**0.66**	**0.65**
$A_{3,Al,D}$	5	0.54	0.60	0.54	0.54
F_{10}	2	**0.83**	**0.83**	**0.83**	**0.83**
$F_{5,Al,S}$	3	**0.79**	**0.79**	**0.79**	**0.79**
$F_{5,Al,D}$	3	**0.79**	0.78	**0.79**	0.78
$F_{4,Al,S}$	4	**0.76**	**0.76**	**0.76**	**0.75**
$F_{4,Al,D}$	4	0.73	0.74	0.73	0.71
$F_{3,Al,S}$	5	**0.75**	**0.76**	**0.75**	**0.75**
$F_{3,Al,D}$	5	0.71	0.72	0.71	0.71

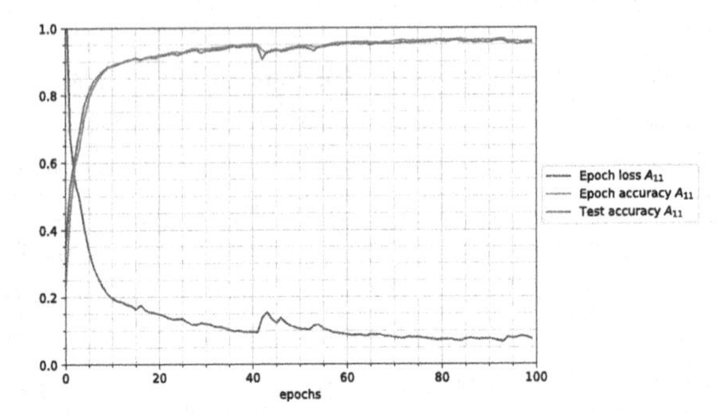

Figure 5.4: Metrics graph of A_{11}. The epoch accuracy tracks the test accuracy.

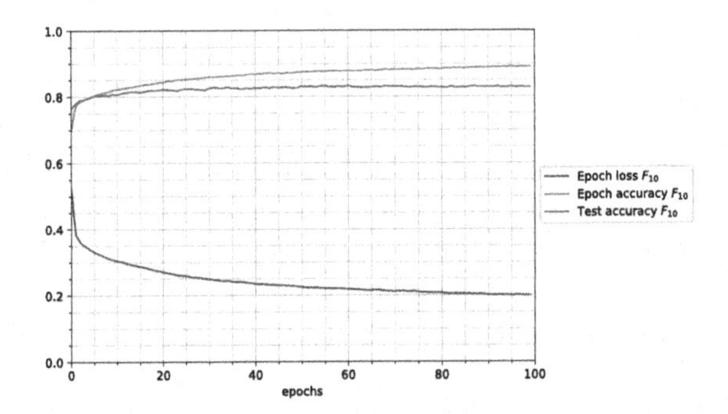

Figure 5.5: Metrics graph of F_{10}. The epoch accuracy increases further while the test accuracy stagnates.

5.3.2 Class Arrangement

Table 5.7 shows the results for the arrangement tests on the Fashion MNIST and SDD datasets. It can be observed that arranging similar classes adjacent to each other outperforms placing distinct classes next to each other. For the Fashion MNIST dataset, the metrics deteriorate further with the maximum number of sectors per neuron. In the SDD case, the metrics with a single output neuron are in fact better than in the $A_{3,AI}$ and $A_{4,AI}$ models. Figure 5.7 shows that for the SDD dataset, training and test accuracy are closely matched over time. For the Fashion MNIST dataset, however, it can be seen in 5.6 that the test accuracies stagnate early on, whereas the training accuracies continue to increase. Since the test accuracy does not decrease, there is no classical overfitting, but the training data are learned considerably more accurately.

Table 5.7: Test results for class arrangement on the Fashion MNIST and AutaASS datasets. The superior result is shown in bold per comparison case. The test cases with similarly classes adjacent outperform the distinct arranged ones.

Model	S/N	Accuracy	Precision	Recall	F1-Score
$A_{1,Arr,S}$	11	**0.83**	**0.83**	**0.83**	**0.83**
$A_{1,Arr,D}$	11	0.62	0.62	0.62	0.61
$F_{1,Arr,S}$	10	**0.67**	**0.67**	**0.67**	**0.67**
$F_{1,Arr,D}$	10	0.57	0.55	0.57	0.55

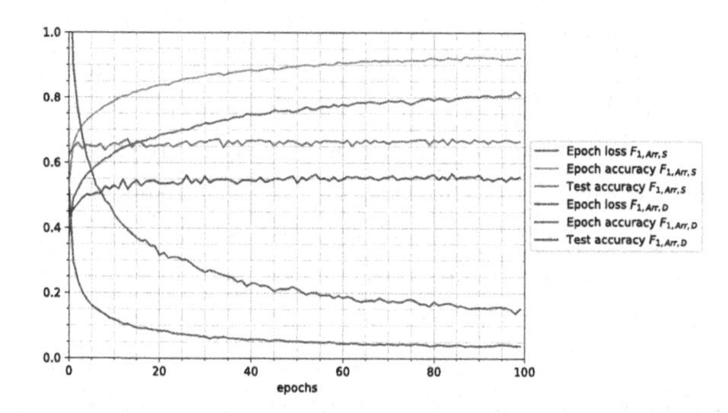

Figure 5.6: Metrics graph of $F_{1,Arr,S}$ and $F_{1,Arr,D}$. The epoch accuracy increases further while the test accuracy stagnates. $F_{1,Arr,S}$ outperforms $F_{1,Arr,D}$.

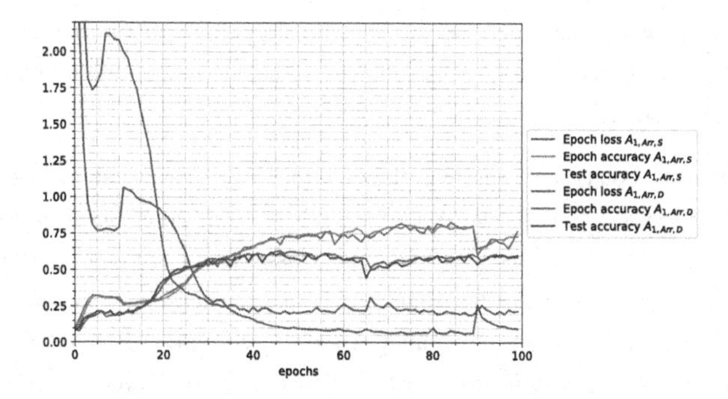

Figure 5.7: Metrics graph of $A_{1,Arr,S}$ and $A_{1,Arr,D}$. The epoch accuracy tracks the test accuracy. $A_{1,Arr,S}$ outperforms $A_{1,Arr,D}$, but learning becomes more unstable.

5.4 Discussion

This section discusses the approach from Chapter 4, evaluating both the preprocessing methods and the novel approach to a class assignment. The results from the previous section serve as a foundation for discussion. This is followed by a general assessment of MLMVNs.

The transformation of the real-valued inputs onto the unit circle in the complex plane by the min-max scaler (4.1) is suitable with the selected datasets. Chapter 4 has already discussed limitations that could affect the application of this method. However, decent results were achieved here without reducing the factor α or performing outlier removal.

The 2D DFT combined with a reduction to low-frequency components may be a reasonable choice to obtain meaningful features and a dimension reduction at the same time. But by mapping to the unit circle, i.e. discarding the magnitude information, the 2D DFT loses essential characteristics such as the translational invariance (4.3)[Bov05]. The MNIST datasets, both the MNIST dataset of handwritten digits and the Fashion MNIST dataset, counteract this by utilizing centered images. Still, even a shift of one pixel results in vastly different phases. In addition, the Fashion MNIST dataset contains more dynamic images in the sense that, for example, two coats are more different from each other than two identical digits. Nevertheless, the method from [AG18] was improved in that the information from the central pixel was additionally extracted. Internal preliminary tests have shown that without this pixel, the accuracy on the Fashion MNIST dataset drops by a ten percentage points. Since this pixel can be interpreted as the average intensity value of the original image, it is evident that especially differently sized objects like sandals and pullovers could be distinguished with this pixel value exclusively. In summary, the method has been improved, but image recognition with the phase information only is inappropriate for sophisticated tasks. The transformation to the frequency domain is a powerful tool in image recognition, but discarding the magnitude information renders MLMVNs unsuitable for this task.

Regarding class assignment, the hypotheses that similar classes should be allocated to the same neuron and that similar classes should be arranged adjacent on the unit circle could be empirically supported. By conceiving classes as a function of their features, it is logical that the weighted sum z for similar input values will produce similar output values. Therefore, arranging distinct and similar classes in alternation is counterproductive. In the allocation process, each neuron must classify all classes. However, this is easier if the neuron only has to distinguish between similar classes and maps the rest of the classes to sector 0. To separate distinct values strongly from sector 0 is a more difficult task.

Nevertheless, the performance metrics decrease with an increasing number of sectors per neuron. While the possibility to map the weighted sum z into the desired sector is infinitely large in terms of magnitude, the valid interval decreases with each additional sector in terms of the possible angular range. Only model $A_{1,Arr,S}$ again increased metrics compared to models with multiple output neurons. This could be explained by the fact that no neurons have to map to the sector class *other*, i.e. sector 0, anymore, but that the output neuron actually indicates all classes. However, this is not a general observation since $F_{1,Arr,S}$ does not demonstrate this behavior. Despite these findings, binary assignment of classes to output neurons has been shown to outperform any other model. Therefore, the assignment approach from Chapter 4 should only be applied when fewer output neurons are available than classes.

In general, shallow MLMVNs with only one hidden layer have been evaluated. These have emerged as the best architectures in both [PL21], and [AG18]. Deep MLMVNs have not been explicitly investigated in more detail, but trial runs with deep architectures have yielded worse results. However, this contrasts with classical RVNNs and CVNNs, where deeper architectures have a higher capacity if they are kept from overfitting. This might be an initialization problem if these MLMVNs did not learn at all or merely took longer to find the optimal weights. With worse results, however, this could be an error backpropagation problem. For MLMVNs, the error is assumed to be uniformly distributed and is shared equally among the neurons of the previous layer. Classical BP, however, employs the partial derivatives with respect to the weights. The extent to which MLMVNs can be deployed with deep architectures remains for future work.

5.5 Summary

This chapter has addressed details of the PyTorch implementation. The two datasets, SDD and Fashion MNIST, were presented and examined for their class similarity. Based on these findings, evaluation tests for the class assignment approach presented in Chapter 4 were created and performed. The transformation according to the min-max scaler was found to be suitable for MLMVNs. The combination of the 2D DFT with MLMVNs has been improved but has revealed systematic weaknesses. The test results supported the hypotheses that similar classes should be allocated to the same neuron and that similar classes should be arranged adjacent to each other. Nevertheless, binary assigned classes outperformed multi-valued ones.

Chapter 6

Conclusion and Outlook

This chapter summarizes the entire thesis and its findings. In addition, an outlook is given on topics that require further consideration.

6.1 Conclusion

Complex-valued neural networks have increased functionality due to their complex weights, biases, and activation functions [Aiz11]. Even with the same number of parameters, they can have performance advantages over their real-valued counterparts depending on the data [TBZ+18]. In this context, MLMVNs are a particular type of complex-valued neural networks that utilize a derivative-free learning algorithm and enable multi-valued decision logic. They divide the unit circle into several sectors to distinguish multiple classes simultaneously.

After introducing MLMVNs, state of the art analysis found that both real-valued and complex-valued networks perform multinomial classification based on softmax function. Even MLMVNs currently leverage their multi-valued decision logic only in architectures with a single output neuron. This raises the issue of the impact of a particular assignment of classes to output neurons on performance. Thereupon, the approach was presented that similar classes should be assigned to the same neuron when using multiple output neurons and that similar classes should be adjacent on the unit circle. Since MLMVNs require input data to be located on the unit circle, two transformations were evaluated. The numerical SDD dataset and the Fashion MNIST image dataset were chosen to assess the methods and the assignment approach. The min-max scaler transformation was found to be suitable for numerical data. However, the 2D DFT has lost properties such as translational invariance due to the lack of magnitude information. As a result, MLMVNs

J. Knaup, *Impact of Class Assignment on Multinomial Classification Using Multi-Valued Neurons*, BestMasters, https://doi.org/10.1007/978-3-658-38955-0_6

with this transformation are not predestined for image recognition. Nevertheless, the postulated hypothesis regarding class assignment could be supported. Regardless, binary-assigned classifications outperformed the novel approach, which means that the proposed approach should only be employed if fewer output neurons than classes are available.

6.2 Outlook

Since the multiple output neurons approach with multiple class assignments was less performant than binary assignment, this approach should not be pursued further. Nevertheless, in general, other points regarding MLMVNs should be investigated further.

The results of this work were obtained with shallow network architectures containing only one hidden layer. In this context, it needs to be clarified whether MLMVNs can implement significantly deeper structures or whether they are limited by their error backpropagation in this respect.

Furthermore, it has been shown that MLMVNs are not primarily suitable for image recognition since classification based only on phase information entails some disadvantages. Another field of application of MLMVNs is image denoising [AOO17]. Here, it should be investigated whether similar limitations in usability are revealed.

The last point refers to the functionality and properties of MLMVNs. First, periodicity for multiple output neurons should be implemented in the current repository to evaluate its impact on multiple binary-assigned neurons. Furthermore, the learning rate is of interest. Compared to real-valued networks, this can also be complex-valued. It should be clarified whether the choice of the learning rate can favorably influence the learning behavior. Moreover, the learning rate could be adapted dynamically depending on the error. Another interesting approach would be using an optimizer analogous to Adam [KB14], for example, where previous error values are included in the weight adjustment. Since the result curves partially show erratic changes, limiting the weight adjustments in a gradient clipping manner might be reasonable.

Appendix A

Results

A.1 Binary Classification

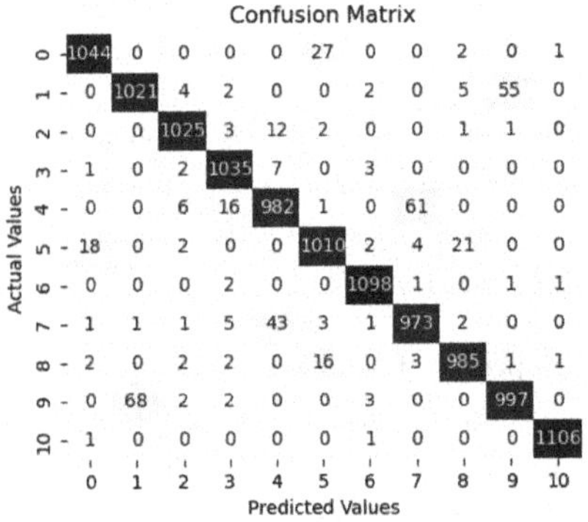

Figure A.1: Confusion matrix of A_{11} with the best-performing weights of the training process.

Class	Precision	Recall	F1-score
0	0.98	0.97	0.98
1	0.94	0.94	0.94
2	0.98	0.98	0.98
3	0.97	0.99	0.98
4	0.94	0.92	0.93
5	0.95	0.96	0.95
6	0.99	1.00	0.99
7	0.93	0.94	0.94
8	0.97	0.97	0.97
9	0.95	0.93	0.94
10	1.00	1.00	1.00
Accuracy			0.96

Table A.1: Metrics of A_{11} with the best-performing weights of the training process.

J. Knaup, *Impact of Class Assignment on Multinomial Classification Using Multi-Valued Neurons*, BestMasters, https://doi.org/10.1007/978-3-658-38955-0

Confusion Matrix

Actual \ Predicted	0	1	2	3	4	5	6	7	8	9
0	767	12	12	42	9	3	135	1	19	0
1	9	946	4	28	6	0	4	0	3	0
2	17	2	718	11	131	2	97	0	22	0
3	49	30	13	843	27	2	22	0	14	0
4	4	1	111	32	753	0	90	0	9	0
5	1	0	0	0	0	909	0	46	12	32
6	136	9	99	46	82	4	599	1	23	1
7	0	0	0	0	0	28	0	912	1	59
8	6	3	10	12	5	14	10	3	933	4
9	0	0	0	0	0	18	0	38	2	942

Actual Values / Predicted Values

Figure A.2: Confusion matrix of F_{10} with the best-performing weights of the training process.

Class	Precision	Recall	F1-score
0	0.78	0.77	0.77
1	0.94	0.95	0.94
2	0.74	0.72	0.73
3	0.83	0.84	0.84
4	0.74	0.75	0.75
5	0.93	0.91	0.92
6	0.63	0.60	0.61
7	0.91	0.91	0.91
8	0.90	0.93	0.92
9	0.91	0.94	0.92
Accuracy			0.83

Table A.2: Metrics of F_{10} with the best-performing weights of the training process.

A.2 Class Allocation

Confusion Matrix

Actual \ Predicted	0	1	2	3	4	5	6	7	8	9	10	11
0	1030	2	0	0	0	0	0	1	28	12	1	0
1	6	1070	0	0	5	11	5	5	0	0	0	1
2	1	0	947	118	5	10	0	0	2	5	1	0
3	2	0	107	951	6	3	0	0	2	0	1	0
4	7	11	3	7	873	110	13	5	5	10	0	0
5	3	6	0	2	52	940	36	8	1	0	0	0
6	0	4	0	0	54	55	790	159	0	4	0	0
7	8	2	1	0	4	3	116	892	0	3	1	0
8	35	6	0	0	6	2	2	11	932	63	0	0
9	4	0	2	3	1	0	3	3	35	960	1	0
10	1	0	0	0	0	0	0	0	0	0	1107	0
11	0	0	0	0	0	0	0	0	0	0	0	0

Actual Values / Predicted Values

Figure A.3: Confusion matrix of $A_{6,Al,S}$ with the best-performing weights of the training process.

Class	Precision	Recall	F1-score
0	0.94	0.96	0.95
1	0.97	0.97	0.97
2	0.89	0.87	0.88
3	0.88	0.89	0.88
4	0.87	0.84	0.85
5	0.83	0.90	0.86
6	0.82	0.74	0.78
7	0.82	0.87	0.84
8	0.93	0.88	0.90
9	0.91	0.95	0.93
10	1.00	1.00	1.00
Accuracy			0.90

Table A.3: Metrics of $A_{6,Al,S}$ with the best-performing weights of the training process.

Confusion Matrix

Actual Values \ Predicted Values	0	1	2	3	4	5	6	7	8	9	10	11
0	1031	4	0	2	0	0	1	0	28	2	5	1
1	2	1033	0	1	1	2	7	2	0	0	0	0
2	0	0	934	27	0	0	24	91	2	4	2	5
3	5	6	39	851	1	2	77	14	7	0	23	5
4	0	3	0	2	1007	9	10	8	2	1	0	2
5	0	0	0	0	0	1101	1	0	0	0	1	0
6	4	13	22	88	6	3	862	54	3	0	8	3
7	0	0	72	11	2	0	46	939	0	0	0	2
8	30	7	4	10	13	1	6	0	866	29	73	18
9	0	0	0	0	0	0	0	1	6	1098	3	0
10	4	5	5	5	1	1	1	4	71	27	843	45
11	0	0	0	0	0	0	0	0	0	0	0	0

Figure A.4: Confusion matrix of $A_{6,Al,D}$ with the best-performing weights of the training process.

Class	Precision	Recall	F1-score
0	0.96	0.96	0.96
1	0.96	0.99	0.97
2	0.87	0.86	0.86
3	0.85	0.83	0.84
4	0.98	0.96	0.97
5	0.98	1.00	0.99
6	0.83	0.81	0.82
7	0.84	0.88	0.86
8	0.88	0.82	0.85
9	0.95	0.99	0.97
10	0.88	0.83	0.86
Accuracy			0.90

Table A.4: Metrics of $A_{6,Al,D}$ with the best-performing weights of the training process.

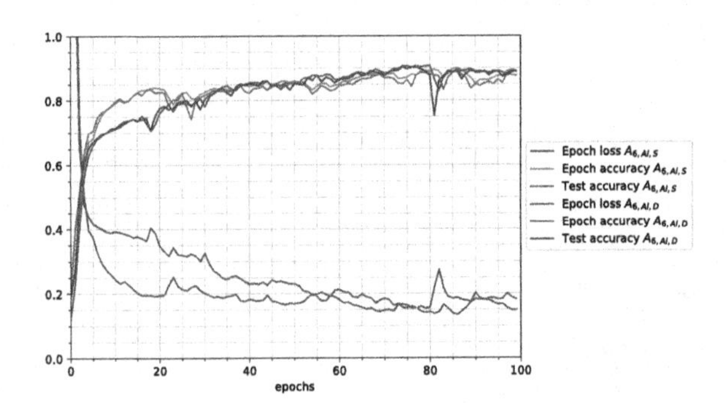

Figure A.5: Metrics graph of $A_{6,Al,S}$ and $A_{6,Al,D}$. The epoch accuracy tracks the test accuracy. Both model performances are close to each other, but $A_{6,Al,S}$ learns faster in the beginning.

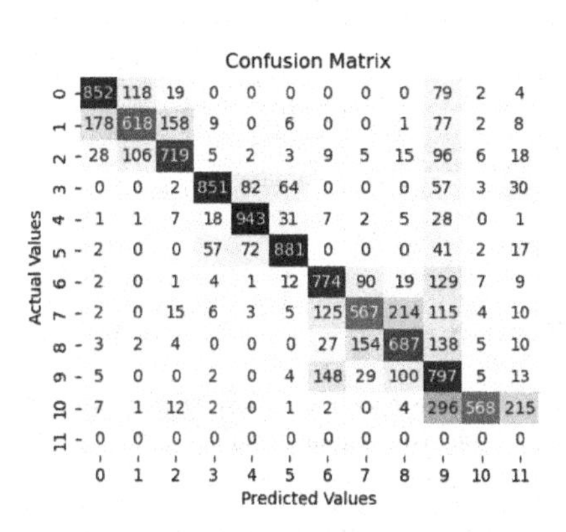

Figure A.6: Confusion matrix of $A_{4,Al,S}$ with the best-performing weights of the training process.

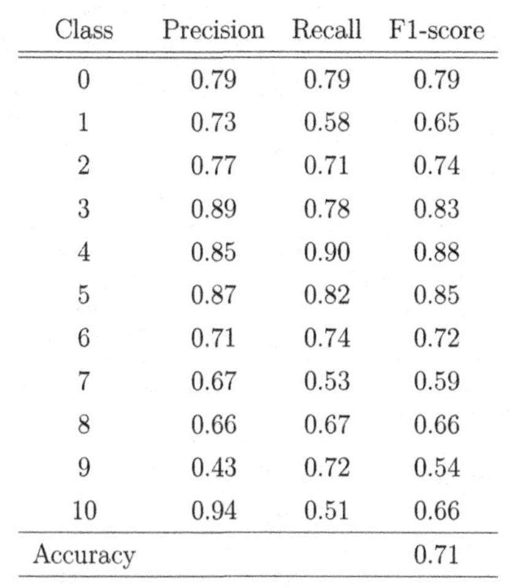

Class	Precision	Recall	F1-score
0	0.79	0.79	0.79
1	0.73	0.58	0.65
2	0.77	0.71	0.74
3	0.89	0.78	0.83
4	0.85	0.90	0.88
5	0.87	0.82	0.85
6	0.71	0.74	0.72
7	0.67	0.53	0.59
8	0.66	0.67	0.66
9	0.43	0.72	0.54
10	0.94	0.51	0.66
Accuracy			0.71

Table A.5: Metrics of $A_{4,Al,S}$ with the best-performing weights of the training process.

Figure A.7: Confusion matrix of $A_{4,Al,D}$ with the best-performing weights of the training process.

Class	Precision	Recall	F1-score
0	0.77	0.80	0.79
1	0.90	0.70	0.79
2	0.82	0.67	0.74
3	0.62	0.82	0.70
4	0.88	0.44	0.59
5	0.90	0.80	0.84
6	0.30	0.59	0.40
7	0.46	0.44	0.45
8	0.60	0.66	0.63
9	0.64	0.59	0.62
10	0.90	0.56	0.69
Accuracy			0.64

Table A.6: Metrics of $A_{4,Al,D}$ with the best-performing weights of the training process.

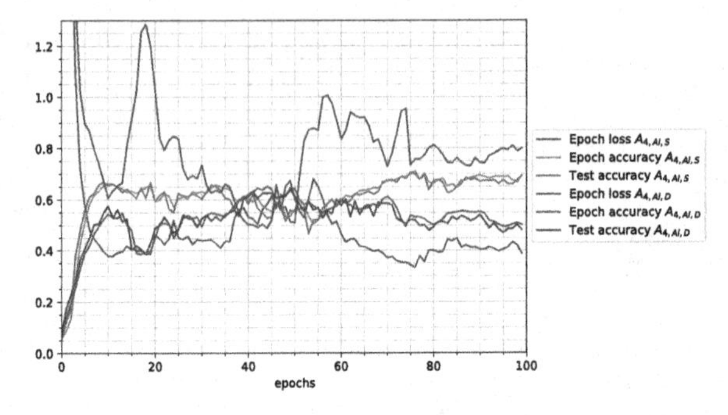

Figure A.8: Metrics graph of $A_{4,Al,S}$ and $A_{4,Al,D}$. The epoch accuracy tracks the test accuracy. $A_{4,Al,S}$ outperforms $A_{4,Al,D}$, but the graphs become erratic.

Confusion Matrix

Actual Values \ Predicted Values	0	1	2	3	4	5	6	7	8	9	10	11
0	770	114	53	32	15	0	0	4	59	3	3	36
1	1	1093	2	1	0	0	0	0	5	0	0	1
2	51	63	776	131	7	0	0	4	29	0	0	11
3	0	0	0	1106	0	0	0	0	1	0	0	1
4	1	0	0	8	839	75	16	17	59	3	1	25
5	2	1	0	1	25	823	164	1	12	1	2	16
6	0	0	0	0	13	146	658	191	30	1	0	27
7	1	0	1	1	11	7	71	832	30	4	4	68
8	5	0	0	13	67	0	0	31	619	109	46	184
9	11	0	0	28	55	0	0	26	574	100	40	223
10	6	0	3	12	53	0	0	48	294	84	51	461
11	0	0	0	0	0	0	0	0	0	0	0	0

Figure A.9: Confusion matrix of $A_{3,Al,S}$ with the best-performing weights of the training process.

Class	Precision	Recall	F1-score
0	0.91	0.71	0.80
1	0.86	0.99	0.92
2	0.93	0.72	0.81
3	0.83	1.00	0.91
4	0.77	0.80	0.79
5	0.78	0.79	0.78
6	0.72	0.62	0.67
7	0.72	0.81	0.76
8	0.36	0.58	0.44
9	0.33	0.09	0.15
10	0.35	0.05	0.09
Accuracy			0.66

Table A.7: Metrics of $A_{3,Al,S}$ with the best-performing weights of the training process.

Confusion Matrix

Actual Values \ Predicted Values	0	1	2	3	4	5	6	7	8	9	10	11
0	1012	22	0	3	2	1	2	12	2	0	0	18
1	90	370	219	29	5	0	0	1	27	19	64	265
2	31	85	512	176	110	66	20	24	19	0	1	22
3	5	2	15	920	8	0	0	3	95	12	3	40
4	44	9	11	53	731	37	0	2	15	2	5	135
5	28	6	18	75	181	610	43	15	39	1	1	13
6	33	2	10	13	22	122	661	93	12	2	1	41
7	2	0	0	0	0	0	8	943	22	13	35	85
8	84	10	13	249	72	0	0	38	351	27	8	196
9	172	12	6	107	194	19	23	143	98	8	5	270
10	66	28	60	94	8	0	0	1	17	35	253	510
11	0	0	0	0	0	0	0	0	0	0	0	0

Figure A.10: Confusion matrix of $A_{3,Al,D}$ with the best-performing weights of the training process.

Class	Precision	Recall	F1-score
0	0.65	0.94	0.77
1	0.68	0.34	0.45
2	0.59	0.48	0.53
3	0.54	0.83	0.65
4	0.55	0.70	0.62
5	0.71	0.59	0.65
6	0.87	0.65	0.75
7	0.74	0.85	0.79
8	0.50	0.33	0.40
9	0.07	0.01	0.01
10	0.67	0.24	0.35
Accuracy			0.54

Table A.8: Metrics of $A_{3,Al,D}$ with the best-performing weights of the training process.

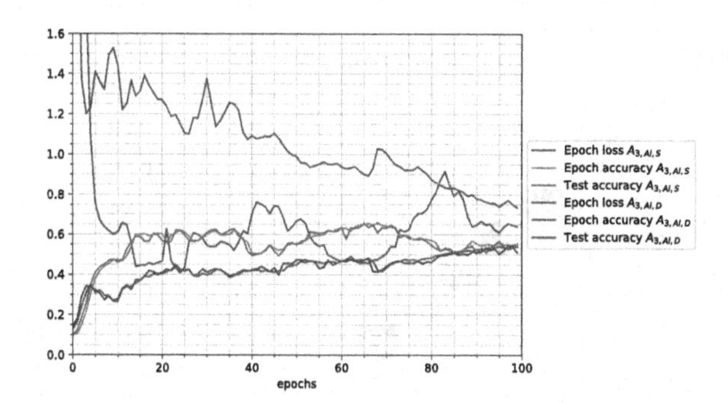

Figure A.11: Metrics graph of $A_{3,Al,S}$ and $A_{3,Al,D}$. The epoch accuracy tracks the test accuracy. $A_{3,Al,S}$ outperforms $A_{3,Al,D}$, but the graphs become erratic.

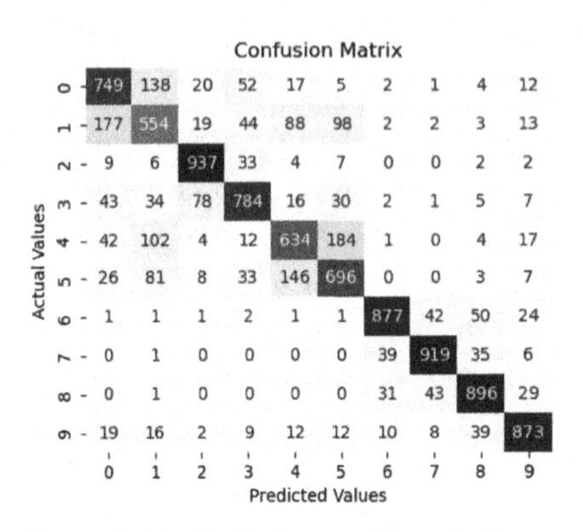

Figure A.12: Confusion matrix of $F_{5,Al,S}$ with the best-performing weights of the training process.

Class	Precision	Recall	F1-score
0	0.70	0.75	0.73
1	0.59	0.55	0.57
2	0.88	0.94	0.91
3	0.81	0.78	0.80
4	0.69	0.63	0.66
5	0.67	0.70	0.68
6	0.91	0.88	0.89
7	0.90	0.92	0.91
8	0.86	0.90	0.88
9	0.88	0.87	0.88
Accuracy			0.79

Table A.9: Metrics of $F_{5,Al,S}$ with the best-performing weights of the training process.

Confusion Matrix

Actual Values \ Predicted Values	0	1	2	3	4	5	6	7	8	9
0	712	39	38	116	17	6	40	13	8	11
1	29	911	0	0	10	33	4	11	1	1
2	8	0	943	17	3	1	20	5	3	0
3	123	26	54	504	102	22	39	14	88	28
4	10	3	21	102	645	47	7	4	119	42
5	4	47	0	0	25	894	9	21	0	0
6	32	6	33	39	9	4	782	48	28	19
7	15	15	0	0	5	37	26	897	3	2
8	3	1	27	63	112	20	35	6	687	46
9	12	11	7	9	11	6	20	11	25	888

Figure A.13: Confusion matrix of $F_{5,Al,D}$ with the best-performing weights of the training process.

Class	Precision	Recall	F1-score
0	0.75	0.71	0.73
1	0.86	0.91	0.88
2	0.84	0.94	0.89
3	0.59	0.50	0.54
4	0.69	0.65	0.67
5	0.84	0.89	0.86
6	0.80	0.78	0.79
7	0.87	0.90	0.88
8	0.71	0.69	0.70
9	0.86	0.89	0.87
Accuracy			0.79

Table A.10: Metrics of $F_{5,Al,D}$ with the best-performing weights of the training process.

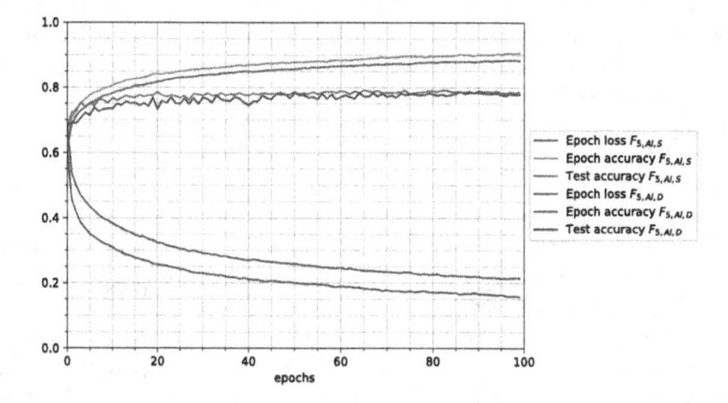

Figure A.14: Metrics graph of $F_{5,Al,S}$ and $F_{5,Al,D}$. The epoch accuracy increases further while the test accuracy stagnates. $F_{5,Al,S}$ outperforms $F_{5,Al,D}$ only on the training data, otherwise they are close to each other.

Confusion Matrix

Actual Values \ Predicted Values	0	1	2	3	4	5	6	7	8	9	10	11
0	723	144	69	25	5	18	1	1	2	8	0	4
1	236	406	124	117	64	47	1	0	1	3	0	1
2	46	68	782	29	16	26	3	0	2	23	2	3
3	79	34	48	656	132	49	2	0	0	0	0	0
4	60	30	49	181	587	88	2	0	1	2	0	0
5	27	7	26	22	44	852	9	4	7	1	0	1
6	2	0	0	4	0	2	880	64	48	0	0	0
7	1	0	3	1	2	4	80	861	48	0	0	0
8	1	0	1	6	0	2	33	54	902	1	0	0
9	11	6	27	3	2	6	0	0	2	929	13	1
10	0	0	0	0	0	0	0	0	0	0	0	0
11	0	0	0	0	0	0	0	0	0	0	0	0

Figure A.15: Confusion matrix of $F_{4,Al,S}$ with the best-performing weights of the training process.

Class	Precision	Recall	F1-score
0	0.61	0.72	0.66
1	0.58	0.41	0.48
2	0.69	0.78	0.73
3	0.63	0.66	0.64
4	0.69	0.59	0.63
5	0.78	0.85	0.81
6	0.87	0.88	0.88
7	0.88	0.86	0.87
8	0.89	0.90	0.90
9	0.96	0.93	0.94
Accuracy			0.76

Table A.11: Metrics of $F_{4,Al,S}$ with the best-performing weights of the training process.

Confusion Matrix (Actual Values × Predicted Values)

Actual \ Predicted	0	1	2	3	4	5	6	7	8	9	10	11
0	869	18	12	50	4	7	20	2	18	0	0	0
1	113	447	130	58	24	53	6	5	19	132	5	8
2	21	26	769	21	6	42	30	13	41	26	2	3
3	29	0	5	887	19	5	40	5	10	0	0	0
4	48	39	71	114	345	120	49	67	62	76	1	8
5	1	1	25	7	8	936	6	4	5	5	0	2
6	18	1	3	35	3	6	911	18	5	0	0	0
7	19	5	40	60	34	65	87	575	106	9	0	0
8	26	5	18	21	2	17	13	19	875	4	0	0
9	58	43	92	41	23	49	3	1	13	664	10	3
10	0	0	0	0	0	0	0	0	0	0	0	0
11	0	0	0	0	0	0	0	0	0	0	0	0

Figure A.16: Confusion matrix of $F_{4,Al,D}$ with the best-performing weights of the training process.

Class	Precision	Recall	F1-score
0	0.72	0.87	0.79
1	0.76	0.45	0.56
2	0.66	0.77	0.71
3	0.69	0.89	0.77
4	0.74	0.34	0.47
5	0.72	0.94	0.81
6	0.78	0.91	0.84
7	0.81	0.57	0.67
8	0.76	0.88	0.81
9	0.72	0.66	0.69
Accuracy			0.73

Table A.12: Metrics of $F_{4,Al,D}$ with the best-performing weights of the training process.

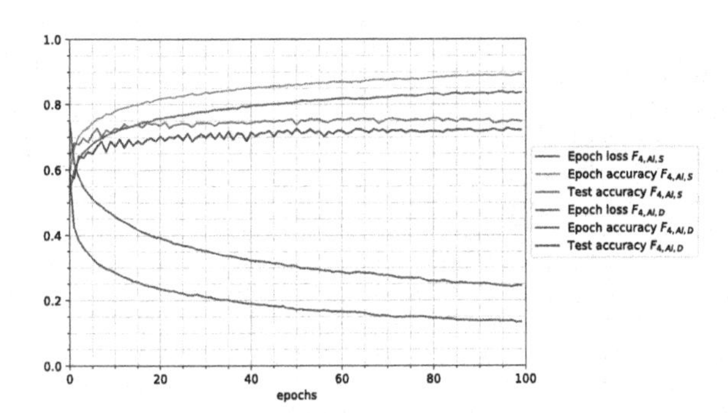

Figure A.17: Metrics graph of $F_{4,Al,S}$ and $F_{4,Al,D}$. The epoch accuracy increases further while the test accuracy stagnates. $F_{4,Al,S}$ outperforms $F_{4,Al,D}$.

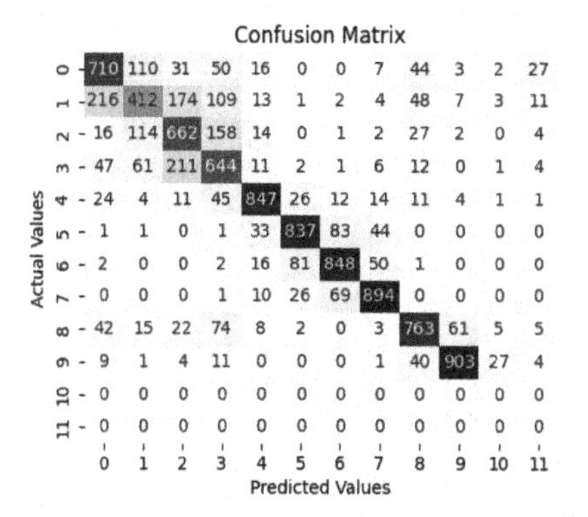

Figure A.18: Confusion matrix of $F_{3,Al,S}$ with the best-performing weights of the training process.

Class	Precision	Recall	F1-score
0	0.67	0.71	0.69
1	0.57	0.41	0.48
2	0.59	0.66	0.63
3	0.59	0.64	0.61
4	0.88	0.85	0.86
5	0.86	0.84	0.85
6	0.83	0.85	0.84
7	0.87	0.89	0.88
8	0.81	0.76	0.78
9	0.92	0.90	0.91
Accuracy			0.75

Table A.13: Metrics of $F_{3,Al,S}$ with the best-performing weights of the training process.

Figure A.19: Confusion matrix of $F_{3,Al,D}$ with the best-performing weights of the training process.

Class	Precision	Recall	F1-score
0	0.73	0.93	0.82
1	0.76	0.66	0.71
2	0.73	0.52	0.61
3	0.67	0.86	0.75
4	0.56	0.67	0.61
5	0.74	0.46	0.57
6	0.84	0.81	0.82
7	0.76	0.85	0.80
8	0.53	0.48	0.50
9	0.91	0.85	0.88
Accuracy			0.71

Table A.14: Metrics of $F_{3,Al,D}$ with the best-performing weights of the training process.

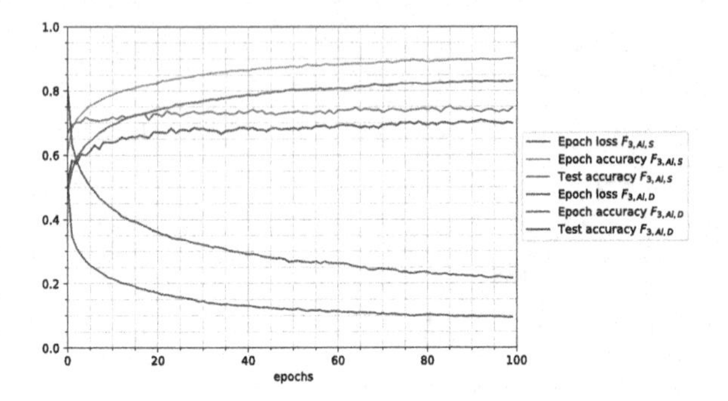

Figure A.20: Metrics graph of $F_{3,Al,S}$ and $F_{3,Al,D}$. The epoch accuracy increases faster than the test accuracy. $F_{3,Al,S}$ outperforms $F_{3,Al,D}$.

A.3 Class Arrangement

Confusion Matrix

Actual Values \ Predicted Values	0	1	2	3	4	5	6	7	8	9	10
0	1071	27	1	0	0	0	0	0	0	0	4
1	26	870	143	6	0	1	1	0	0	0	1
2	0	28	830	184	2	0	0	0	0	0	0
3	0	6	147	720	187	6	0	0	0	0	0
4	1	2	4	162	821	40	0	0	0	0	0
5	0	0	0	0	14	926	132	1	0	0	1
6	0	0	1	0	1	120	853	79	2	0	1
7	1	1	0	0	0	1	103	832	68	6	0
8	1	0	0	0	0	1	2	38	862	180	5
9	0	0	0	0	0	0	0	5	171	851	45
10	10	1	0	0	0	0	0	0	0	35	1062

Figure A.21: Confusion matrix of $A_{1,Arr,S}$ with the best-performing weights of the training process.

Class	Precision	Recall	F1-score
0	0.96	0.97	0.97
1	0.93	0.83	0.88
2	0.74	0.80	0.76
3	0.67	0.68	0.67
4	0.80	0.80	0.80
5	0.85	0.86	0.85
6	0.78	0.81	0.79
7	0.87	0.82	0.85
8	0.78	0.79	0.79
9	0.79	0.79	0.79
10	0.95	0.96	0.95
Accuracy			0.83

Table A.15: Metrics of $A_{1,Arr,S}$ with the best-performing weights of the training process.

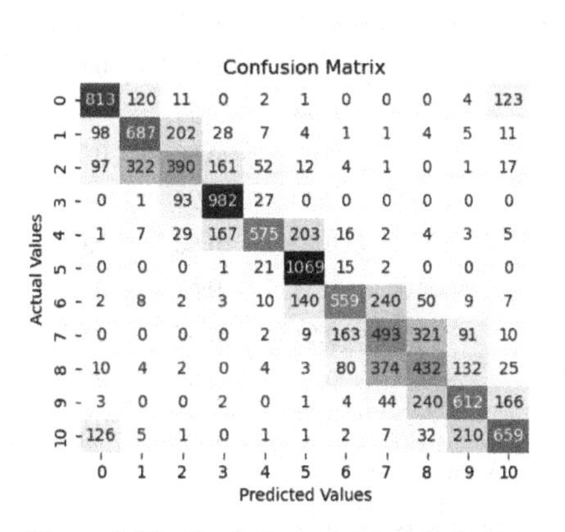

Figure A.22: Confusion matrix of $A_{1,Arr,D}$ with the best-performing weights of the training process.

Class	Precision	Recall	F1-score
0	0.71	0.76	0.73
1	0.60	0.66	0.62
2	0.53	0.37	0.44
3	0.73	0.89	0.80
4	0.82	0.57	0.67
5	0.74	0.96	0.84
6	0.66	0.54	0.60
7	0.42	0.45	0.44
8	0.40	0.41	0.40
9	0.57	0.57	0.57
10	0.64	0.63	0.64
Accuracy			0.62

Table A.16: Metrics of $A_{1,Arr,D}$ with the best-performing weights of the training process.

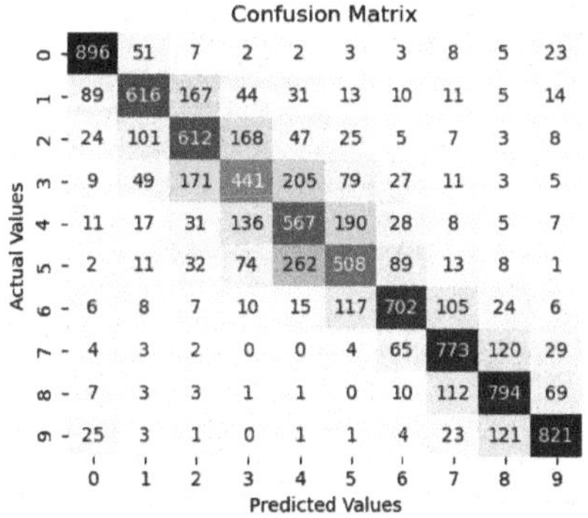

Figure A.23: Confusion matrix of $F_{1,Arr,S}$ with the best-performing weights of the training process.

Class	Precision	Recall	F1-score
0	0.84	0.90	0.86
1	0.71	0.62	0.66
2	0.59	0.61	0.60
3	0.50	0.44	0.47
4	0.50	0.57	0.53
5	0.54	0.51	0.52
6	0.74	0.70	0.72
7	0.72	0.77	0.75
8	0.73	0.79	0.76
9	0.84	0.82	0.83
Accuracy			0.67

Table A.17: Metrics of $F_{1,Arr,S}$ with the best-performing weights of the training process.

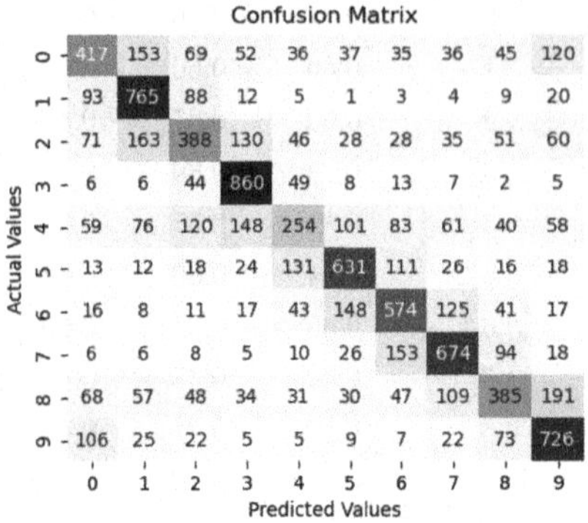

Confusion Matrix

Actual \ Predicted	0	1	2	3	4	5	6	7	8	9
0	417	153	69	52	36	37	35	36	45	120
1	93	765	88	12	5	1	3	4	9	20
2	71	163	388	130	46	28	28	35	51	60
3	6	6	44	860	49	8	13	7	2	5
4	59	76	120	148	254	101	83	61	40	58
5	13	12	18	24	131	631	111	26	16	18
6	16	8	11	17	43	148	574	125	41	17
7	6	6	8	5	10	26	153	674	94	18
8	68	57	48	34	31	30	47	109	385	191
9	106	25	22	5	5	9	7	22	73	726

Actual Values / Predicted Values

Figure A.24: Confusion matrix of $F_{1,Arr,D}$ with the best-performing weights of the training process.

Class	Precision	Recall	F1-score
0	0.49	0.42	0.45
1	0.60	0.77	0.67
2	0.48	0.39	0.43
3	0.67	0.86	0.75
4	0.42	0.25	0.32
5	0.62	0.63	0.63
6	0.54	0.57	0.56
7	0.61	0.67	0.64
8	0.51	0.39	0.44
9	0.59	0.73	0.65
Accuracy			0.57

Table A.18: Metrics of $F_{1,Arr,D}$ with the best-performing weights of the training process.

Bibliography

[AA14] E. Aizenberg and I. Aizenberg. Batch linear least squares-based learning algorithm for MLMVN with soft margins. In *2014 IEEE Symposium on Computational Intelligence and Data Mining (CIDM)*, pages 48–55, 2014.

[AAANM11] F. Amin, M. Amin, A. Y. H. Al Nuaimi, and K. Murase. Wirtinger Calculus Based Gradient Descent and Levenberg-Marquardt Learning Algorithms in Complex-Valued Neural Networks. In *18th International Conference on Neural Information Processing*, volume 7062, pages 550–559. Springer, 11 2011.

[AABF00] I. Aizenberg, N. Aizenberg, C. Butakov, and E. Farberov. Image Recognition on the Neural Network based on Multi-Valued Neurons. In *Proceedings 15th International Conference on Pattern Recognition. ICPR-2000*, volume 2, pages 989–992 vol.2, 2000.

[AG18] I. Aizenberg and A. Gonzalez. Image Recognition using MLMVN and Frequency Domain Features. In *2018 International Joint Conference on Neural Networks (IJCNN)*, pages 1–8, 2018.

[AIP71] N. N. Aizenberg, Y. L. Ivaskiv, and D. A. Pospelov. A certain Generalization of the Threshold Functions. *The Reports of the Acadamy of Sciences of the USSR*, 196:1287–1290, 1971.

[AIPK71] N. N. Aizenberg, Y. L. Ivas'kiv, D. A. Pospelov, and Khudyakov. Multi-Valued Threshold Functions. I. Boolean Complex-Threshold Functions and their Generalization. *Cybernetics*, 4:1287–1290, 1971.

[Aiz11] I. Aizenberg. *Complex-Valued Neural Networks with Multi-Valued Neurons.* Studies in Computational Intelligence. Springer-Verlag, Berlin Heidelberg, 1st ed. edition, 2011.

[Aiz14] I. Aizenberg. MLMVN With Soft Margins Learning. *IEEE Transactions on Neural Networks and Learning Systems*, 25(9):1632–1644, 2014.

J. Knaup, *Impact of Class Assignment on Multinomial Classification Using Multi-Valued Neurons*, BestMasters, https://doi.org/10.1007/978-3-658-38955-0

[ALM12] I. Aizenberg, A. Luchetta, and S. Manetti. A modified learning algorithm for the multilayer neural network with multi-valued neurons based on the complex QR decomposition. *Soft Computing*, 16(4):563–575, 2012.

[Alp14] E. Alpaydin. *Introduction to Machine Learning*. Adaptive Computation and Machine Learning. MIT Press, Cambridge, MA, 3 edition, 2014.

[AOO17] I. Aizenberg, A. Ordukhanov, and F. O'Boy. MLMVN as an intelligent image filter. *2017 International Joint Conference on Neural Networks (IJCNN)*, pages 3106–3113, 2017.

[AR75] N. Ahmed and K.R. Rao. *Orthogonal Transforms for Digital Signal Processing*. Springer-Verlag, Berlin Heidelberg, 1st ed. edition, 1975.

[BB08] W. Burger and M. J. Burge. *Digital Image Processing - An Algorithmic Introduction using Java*. Texts in Computer Science. Springer, 2008.

[BDML12] M. Bator, A. Dicks, U. Mönks, and V. Lohweg. Feature Extraction and Reduction Applied to Sensorless Drive Diagnosis. In *22nd Workshop Computational Intelligence (VDI/VDE-Gesellschaft Mess- und Automatisierungstechnik (GMA))*, 12 2012.

[BERB+13] C. Bayer, O. Enge-Rosenblatt, M. Bator, U. Mönks, A. Dicks, and V. Lohweg. Sensorless drive diagnosis using automated feature extraction, significance ranking and reduction. In *18th IEEE Int. Conf. on Emerging Technologies and Factory Automation (ETFA 2013): IEEE*, pages 1–4, 09 2013.

[BHX+22] A. Baevski, W. Hsu, Q. Xu, A. Babu, J. Gu, and M. Auli. data2vec: A General Framework for Self-supervised Learning in Speech, Vision and Language. January 2022.

[Bis06] C. M. Bishop. *Pattern Recognition and Machine Learning*. Information Science and Statistics. Springer-Verlag, New York, 1st ed. edition, 2006.

[BMR+20] T. Brown, B. Mann, N. Ryder, M. Subbiah, J. D Kaplan, P. Dhariwal, A. Neelakantan, P. Shyam, G. Sastry, A. Askell, S. Agarwal, A. Herbert-Voss, G. Krueger, T. Henighan, R. Child, A. Ramesh, D. Ziegler, J. Wu, C. Winter, C. Hesse, M. Chen, E. Sigler, M. Litwin, S. Gray, B. Chess, J. Clark, C. Berner, S. McCandlish, A. Radford, I. Sutskever, and D. Amodei. Language Models are Few-Shot Learners. In H. Larochelle, M. Ranzato, R. Hadsell, M. F. Balcan, and H. Lin, editors, *Advances in*

Neural Information Processing Systems, volume 33, pages 1877–1901. Curran Associates, Inc., 2020.

[Bov05] A. C. Bovik. *Handbook of Image and Video Processing.* Elsevier Academic Press, USA, 2nd ed. edition, 2005.

[BP92] N. Benvenuto and F. Piazza. On the complex backpropagation algorithm. *IEEE Transactions on Signal Processing*, 40(4):967–969, 1992.

[BQL21] J. Bassey, L Qian, and X. Li. A Survey of Complex-Valued Neural Networks, 2021.

[BRM+21] J. A. Barrachina, C. Ren, C. Morisseau, G. Vieillard, and J. Ovarlez. Complex-Valued vs. Real-Valued Neural Networks for Classification Perspectives: An Example on Non-Circular Data, 2021.

[BSF94] Y. Bengio, P. Simard, and P. Frasconi. Learning long-term dependencies with gradient descent is difficult. *IEEE Transactions on Neural Networks*, 5(2):157–166, 1994.

[BSMM15] I. N. Bronshtein, K.A Semendyayev, G. Musiol, and H. Mühlig. *Handbook of mathematics.* Springer, Berlin, 6th ed. edition, 2015.

[Cau25] A. L. Cauchy. *Mémoire sur les intégrales définies - prises entre des limites imaginaires.* De Bure frères, Paris, 1825.

[CMS12] D. C. Ciresan, U. Meier, and J. Schmidhuber. Multi-column Deep Neural Networks for Image Classification. *CoRR*, abs/1202.2745, 2012.

[Cre93] D. Crevier. *AI: The Tumultuous History Of The Search For Artificial Intelligence.* Basic Books, 1993.

[Dal18] H. Dalianis. *Evaluation Metrics and Evaluation*, pages 45–53. Springer International Publishing, Cham, 2018.

[DDS+09] J. Deng, W. Dong, R. Socher, L. Li, K. Li, and F. Li. ImageNet: A large-scale hierarchical image database. In *2009 IEEE Conference on Computer Vision and Pattern Recognition*, pages 248–255, 2009.

[DG17] D. Dua and C. Graff. UCI Machine Learning Repository, 2017.

[DL14] H. Dörksen and V. Lohweg. Combinatorial refinement of feature weighting for linear classification. In *Proceedings of the 2014 IEEE Emerging Technology and Factory Automation (ETFA)*, pages 1–7, 2014.

[DLLT21] Z. Dai, H Liu, Q. V. Le, and M. Tan. CoAtNet: Marrying Convolution and Attention for All Data Sizes, 2021.

[Fis36] R. A. Fisher. The Use of Multiple Measurements in Taxonomic Problems. *Annals of Eugenics*, 7(2):179–188, 1936.

[Fis02] R. Fischer. *Precoding and signal shaping for digital transmission.* John Wiley & Sons, 2002.

[GB10] X. Glorot and Y. Bengio. Understanding the difficulty of training deep feedforward neural networks. In Yee Whye Teh and Mike Titterington, editors, *Proceedings of the Thirteenth International Conference on Artificial Intelligence and Statistics*, volume 9 of *Proceedings of Machine Learning Research*, pages 249–256, Chia Laguna Resort, Sardinia, Italy, 13–15 May 2010. PMLR.

[GBC16] I. J. Goodfellow, Y. Bengio, and A. Courville. *Deep Learning.* MIT Press, Cambridge, MA, USA, 2016. `http://www.deeplearningbook.org`.

[Gla21] A. Glassner. *Deep Learning: A Visual Approach.* No Starch Press, 2021.

[Heb49] D. O. Hebb. *The organization of behavior: A neuropsychological theory.* Wiley, New York, June 1949.

[Hir12] A. Hirose. *Complex-Valued Neural Networks.* Studies in Computational Intelligence. Springer-Verlag, Berlin Heidelberg, 2nd ed. edition, 2012.

[Hog07] Leslie Hogben, editor. *Handbook of Linear Algebra.* CRC Press, Boca Raton, FL, USA, 2007.

[HS97] S. Hochreiter and J. Schmidhuber. Long Short-term Memory. *Neural computation*, 9(8):1735–1780, 1997.

[HSL+98] N. E. Huang, Z. Shen, S. R. Long, M. C. Wu, H. H. Shih, Q. Zheng, N. C. Yen, C. C. Tung, and H. H. Liu. The empirical mode decomposition and the Hilbert spectrum for nonlinear and non-stationary time series analysis. *Proceedings of the Royal Society of London Series A*, 454(1971):903–998, March 1998.

[HZRS15] K. He, X. Zhang, S. Ren, and J. Sun. Delving Deep into Rectifiers: Surpassing Human-Level Performance on ImageNet Classification. In *2015 IEEE International Conference on Computer Vision (ICCV)*, pages 1026–1034, 2015.

[IS15] S. Ioffe and C. Szegedy. Batch Normalization: Accelerating Deep Network Training by Reducing Internal Covariate Shift. *CoRR*, abs/1502.03167, 2015.

[KB14] D. P. Kingma and J. Ba. Adam: A Method for Stochastic Optimization, 2014. cite arxiv:1412.6980Comment: Published as a conference paper at the 3rd International Conference for Learning Representations, San Diego, 2015.

[KSH12] A. Krizhevsky, I. Sutskever, and G. E. Hinton. ImageNet Classification with Deep Convolutional Neural Networks. In F. Pereira, C. J. C. Burges, L. Bottou, and K. Q. Weinberger, editors, *Advances in Neural Information Processing Systems*, volume 25. Curran Associates, Inc., 2012.

[LC10] Y. LeCun and C. Cortes. MNIST handwritten digit database. 2010.

[Lio] M. Liouville. Leçons sur les fonctions doublement périodiques faites en 1847 par M. J. Liouville*). *Journal für die reine und angewandte Mathematik (Crelles Journal)*, 1880:277 – 310.

[MHM18] L. McInnes, J. Healy, and J. Melville. UMAP: Uniform Manifold Approximation and Projection for Dimension Reduction. *ArXiv e-prints*, February 2018.

[MHSG18] L. McInnes, J. Healy, N. Saul, and L. Grossberger. UMAP: Uniform Manifold Approximation and Projection. *The Journal of Open Source Software*, 3(29):861, 2018.

[Mit97] T. M. Mitchell. *Machine Learning*. McGraw-Hill, New York, 1997.

[MM18] N. Mönning and S. Manandhar. Evaluation of Complex-Valued Neural Networks on Real-Valued Classification Tasks. *CoRR*, abs/1811.12351, 2018.

[MMS93] M. P. Marcus, M. A. Marcinkiewicz, and B. Santorini. Building a Large Annotated Corpus of English: The Penn Treebank. *Comput. Linguist.*, 19(2):313–330, jun 1993.

[MP43] W. Mcculloch and W. Pitts. A Logical Calculus of Ideas Immanent in Nervous Activity. *Bulletin of Mathematical Biophysics*, 5:127–147, 1943.

[MP69] M. Minsky and S. Papert. *Perceptrons: An Introduction to Computational Geometry*. MIT Press, Cambridge, MA, USA, 1969.

[NH10] V. Nair and G. E. Hinton. Rectified Linear Units Improve Restricted Boltzmann Machines. In *Proceedings of the 27th International Conference on International Conference on Machine Learning*, ICML'10, page 807–814, Madison, WI, USA, 2010. Omnipress.

[OL81] A.V. Oppenheim and J.S. Lim. The importance of phase in signals. *Proceedings of the IEEE*, 69(5):529–541, 1981.

[Pea01] K. Pearson. LIII. On lines and planes of closest fit to systems of points in space. *The London, Edinburgh, and Dublin Philosophical Magazine and Journal of Science*, 2(11):559–572, 1901.

[PGM+19] A. Paszke, S. Gross, F. Massa, A. Lerer, J. Bradbury, G. Chanan, T. Killeen, Z. Lin, N. Gimelshein, L. Antiga, A. Desmaison, A. Köpf, E. Yang, Z. DeVito, M. Raison, A. Tejani, S. Chilamkurthy, B. Steiner, L. Fang, J. Bai, and S. Chintala. PyTorch: An Imperative Style, High-Performance Deep Learning Library. *CoRR*, abs/1912.01703, 2019.

[PL21] A. Pfeifer and V. Lohweg. Classification of Faults in Cyber-Physical Systems with Complex-Valued Neural Networks. In *26th IEEE International Conference on Emerging Technologies and Factory Automation (ETFA)*, Västerås, Schweden, Sep 2021. IEEE.

[PVG+11] F. Pedregosa, G. Varoquaux, A. Gramfort, V. Michel, B. Thirion, O. Grisel, M. Blondel, P. Prettenhofer, R. Weiss, V. Dubourg, J. Vanderplas, A. Passos, D. Cournapeau, M. Brucher, M. Perrot, and E. Duchesnay. Scikit-learn: Machine Learning in Python. *Journal of Machine Learning Research*, 12:2825–2830, 2011.

[RHW86] D. E. Rumelhart, G. E. Hinton, and R. J. Williams. Learning Representations by Back-propagating Errors. *Nature*, 323(6088):533–536, 1986.

[RN10] S. Russell and P. Norvig. *Artificial Intelligence: A Modern Approach.* Prentice Hall, 3 edition, 2010.

[Ros58] F. Rosenblatt. The perceptron: A probabilistic model for information storage and organization in the brain. *Psychological Review*, 65(6):386–408, 1958.

[RSG16] M. T. Ribeiro, S. Singh, and C. Guestrin. "Why Should I Trust You?": Explaining the Predictions of Any Classifier. In *Proceedings of the 22nd ACM SIGKDD International Conference on Knowledge Discovery and Data*

Mining, KDD '16, page 1135–1144, New York, NY, USA, 2016. Association for Computing Machinery.

[Sch14] J. Schmidhuber. Deep Learning in Neural Networks: An Overview. *CoRR*, abs/1404.7828, 2014.

[Sha84] C. E. Shannon. Communication in the presence of noise. In *Proceedings of the IEEE*, volume 72, pages 1192–1201, 1984.

[SHK+14] N. Srivastava, G. Hinton, A. Krizhevsky, I. Sutskever, and R. Salakhutdinov. Dropout: A Simple Way to Prevent Neural Networks from Overfitting. *Journal of Machine Learning Research*, 15(56):1929–1958, 2014.

[SHM+16] D. Silver, A. Huang, C. J. Maddison, A. Guez, L. Sifre, G. van den Driessche, J. Schrittwieser, I. Antonoglou, V. Panneershelvam, M. Lanctot, S. Dieleman, D. Grewe, J. Nham, N. Kalchbrenner, I. Sutskever, T. Lillicrap, M. Leach, K. Kavukcuoglu, T. Graepel, and D. Hassabis. Mastering the Game of Go with Deep Neural Networks and Tree Search. *Nature*, 529(7587):484–489, January 2016.

[SL09] M. Sokolova and G. Lapalme. A systematic analysis of performance measures for classification tasks. *Information Processing & Management*, 45(4):427–437, 2009.

[SVVHU18] S. Scardapane, S. Van Vaerenbergh, A. Hussain, and A. Uncini. Complex-valued Neural Networks with Non-parametric Activation Functions. *CoRR*, abs/1802.08026, 2018.

[Sza21] T. Szandała. *Review and Comparison of Commonly Used Activation Functions for Deep Neural Networks*, pages 203–224. Springer Singapore, Singapore, 2021.

[TBZ+18] C. Trabelsi, O. Bilaniuk, Y. Zhang, D. Serdyuk, S. Subramanian, J. F. Santos, S. Mehri, N. Rostamzadeh, Y. Bengio, and C. J. Pal. Deep Complex Networks. *International Conference on Learning Representations (ICLR)*, 2018.

[TKK20] M. S. Tanveer, M. U. K. Khan, and C. Kyung. Fine-Tuning DARTS for Image Classification. *CoRR*, abs/2006.09042, 2020.

[vdMH08] L. van der Maaten and G. Hinton. Visualizing Data using t-SNE. *Journal of Machine Learning Research*, 9(86):2579–2605, 2008.

[Wer74] P. J. Werbos. *Beyond Regression: New Tools for Prediction and Analysis in the Behavioral Sciences.* PhD thesis, Harvard University, 1974.

[WH60] B. Widrow and M. E. Hoff. Adaptive Switching Circuits. In *1960 IRE WESCON Convention Record, Part 4*, pages 96–104, New York, 1960. IRE.

[Wir27] W. Wirtinger. Zur formalen Theorie der Funktionen von mehr komplexen Veränderlichen. *Mathematische Annalen*, 97:357–376, 1927.

[XRV17] H. Xiao, K. Rasul, and R. Vollgraf. Fashion-MNIST: a Novel Image Dataset for Benchmarking Machine Learning Algorithms. *CoRR*, abs/1708.07747, 2017.

[Yud08] E. Yudkowsky. Artificial Intelligence as a Positive and Negative Factor in Global Risk. In Nick Bostrom and Milan Ćirković, editors, *Global Catastrophic Risks*, chapter 15, pages 308–345. Oxford University Press, Oxford, UK, 2008.

[ZLH+12] D. Zimmer, C. Lessmeier, K. Hameyer, C. Piantsop Mbo'o, and I. Coenen. Untersuchung von Bauteilschäden elektrischer Antriebsstränge im Belastungsprüfstand mittels Statorstromanalyse. *ant Journal - Anwendungsnahe Forschung für Antriebstechnik im Maschinenbau*, 1:8–13, 2012.

List of Figures

J. Knaup, *Impact of Class Assignment on Multinomial Classification Using Multi-Valued Neurons*, BestMasters, https://doi.org/10.1007/978-3-658-38955-0

List of Tables

J. Knaup, *Impact of Class Assignment on Multinomial Classification Using Multi-Valued Neurons*, BestMasters, https://doi.org/10.1007/978-3-658-38955-0

Printed in the United States
by Baker & Taylor Publisher Services